Pelican Books

Investigative Design and Statistics

George Wright was born in 1952. He now works as Senior Lecturer in the Psychology Department at the City of London Polytechnic, where he is also Director of the Decision Analysis Group. He currently directs an Economic and Social Research Council project concerned with judgements of the likelihood of future events. Dr Wright received his Ph.D. from Brunel University in 1980 and has since published widely on the human aspects of decision-making. He has also written *Behavioural Decision Theory: An Introduction*, published by Penguin in 1984.

Chris Fowler was born in 1952 and is now Senior Lecturer in Psychology in the Department of Behavioural Sciences, Huddersfield Polytechnic. He presently teaches cognitive psychology and research methods, and prior to this he lectured in psychology and statistics at the University of Nigeria for two years. Dr Fowler read for a joint honours degree in Psychology and Sociology at Birmingham University, and subsequently obtained a Ph.D. in Psychology from Bedford College, University of London, in 1977. His main research interests are in the areas of cross-cultural psychology and human cognition, with particular reference to individual differences and their effects on both human–computer interaction and memory.

George Wright and Chris Fowler

Investigative Design and Statistics

Penguin Books

Penguin Books Ltd, Harmondsworth, Middlesex, England
Viking Penguin Inc., 40 West 23rd Street, New York, New York 10010, U.S.A.
Penguin Books Australia Ltd, Ringwood, Victoria, Australia
Penguin Books Canada Ltd, 2801 John Street, Markham, Ontario, Canada L3R 1B4
Penguin Books (N.Z.) Ltd, 182–190 Wairau Road, Auckland 10, New Zealand

First published 1986

Made and printed in Great Britain by
Cox and Wyman Ltd, Reading
Typeset in Linotron Trump Medieval by
Rowland Phototypesetting Ltd,
Bury St Edmunds, Suffolk

FOR JO AND JOSEPHINE

Contents

Acknowledgements

This book is very much an equal effort by both authors. Ordering of the authors is random. George Wright has primary responsibility for Chapters 4, 6, 7, 8, 11 and 12. Chris Fowler is responsible for Chapters 1, 2, 3, 5, 9 and 10. As will become clear when you read through the book, Chris Fowler is the experimental psychologist and George Wright is the correlational psychologist!

The polished state of this book is due to several people. Jo Fowler's efforts proved invaluable in making the theoretical chapters more comprehensible and readable. We would also like to thank Colin Robson for his patience, invaluable advice and help, and Alastair McClelland, John Wilding, Larry Phillips, Peter Whalley, Peter Ayrton and Veronica Laxon for their detailed comments, particularly on the statistical side of the book. In addition, we would like to thank our respective colleagues at City of London Polytechnic and Huddersfield Polytechnic who offered us support, encouragement and useful suggestions regarding the contents of our earlier drafts. We have incorporated most of these comments and criticisms but the final responsibility for the book remains with us, the authors.

Introduction

Our book is designed as a companion volume to Colin Robson's *Experiment, Design and Statistics in Psychology*. Its aim is to extend introductory coverage of investigative design and statistics to the advanced levels present in the second and third years of an honours degree in Psychology. The book is concerned with the two main research formats used to collect data: the experiment and the correlational study. Whilst the experiment is the basic paradigm for research in experimental and cognitive psychology, the correlational study is the basis for research into individual differences, personality and cross-cultural psychology.

In the earlier chapters, we set the scene for later discussions about designing, carrying out, analysing and interpreting the results obtained from these two investigative approaches. Chapter 1 describes the various processes and stages involved in doing research. Psychological research is characterized as a decision-making process. Next, we devote two chapters to those factors (social and political, and measurement and statistics) which tend to influence these decisions.

In Chapters 4 and 5 we bridge these earlier chapters with later, more statistical chapters. Chapter 4 is devoted to an analysis of the advantages and limitations of the experimental and correlational research formats. In Chapter 5, we review the basic concepts and assumptions underlying statistical tests in general. In particular, we pay attention to the distinction between parametric and non-parametric statistics.

The statistical content of the volume is based on the same non-mathematical 'cookbook' approach as Colin Robson's, in that we provide a fully worked example and significance table for each type of test discussed. The results and procedures of 'typical' experimental/correlational research are used as scenarios for the worked examples. However, before presenting a worked example we give a *verbal* exposition of the statistical assumptions and underpinnings of each test. In particular, emphasis is placed on the location of causation in experimental and correlational investigations and also on the inferences that can be drawn from a particular test result.

Chapters 6, 7 and 8 are concerned with the statistical techniques underlying the correlational research format. Chapter 6 deals with the measurement of association and correlation, whilst Chapter 7 focuses on regression analysis. These two chapters provide a background for Chapter 8, which discusses questionnaire design and analysis. Here we provide a grounding in the theory and statistics necessary to enable you to construct and evaluate your own questionnaires.

The next two chapters, 9 and 10, are devoted to analysis of variance techniques, the major statistical procedures used in the experimental research format. Analysis of variance allows you to evaluate the effects and interaction of several independent variables on a single dependent variable.

In Chapter 11 we provide a *qualitative* discussion of the basis of more advanced multivariate techniques. We include discussion of partial correlation, multiple regression, factor analysis and multidimensional scaling. As statistical knowledge in part determines research design, this chapter is intended to act as a method of alerting you to the further possibilities of investigative design using computer-aided data analysis for the more complex calculations.

Finally, in Chapter 12, we attempt to reconcile the experimental and correlational research formats by adopting two approaches in the discussion. First, we adopt a statistical approach, by showing that a problem involving the comparison of two means can be converted to a problem of estimating the strength of association between two variables. Second, we present an empirical approach by showing the usefulness of analysis of variance methodology in the domain of personality psychology.

1 The Process of Psychological Research

In this chapter we will concentrate on the major issues of why we carry out research, and the nature and importance of the research process itself. In addition, we will introduce many of the issues, questions and concepts discussed in greater depth later in the book.

The Research Process

1 Introduction

Psychology is founded on a scientific approach to the gathering and interpretation of facts on which we base our understanding of various aspects of behaviour.

The carrying out of a research project can be seen as a process which reflects this general approach. It aims to gather precise data through a method which will allow others to replicate (or repeat) particular investigations.

However, research cannot be viewed simply as the gathering and recording of information or data. The data are often collected and gathered within the confines of specific aims. We are guided in the ways in which we collect, analyse and interpret data by the adoption of a scientific approach in general, and by the particular rules and procedures peculiar to the individual discipline, such as psychology. Research carried out in this way ensures that the data collected and subsequent conclusions will be accepted as valid and objective by most other researchers.

Research can be regarded as a process which involves a progression through a number of interdependent stages by means of actions based on decisions. By adopting a holistic approach to the discussion of psychological research, we hope to satisfy two major objectives. First, we hope to avoid the often unintentional compartmentalization of the various stages of research, which can frequently give the impression that they are independent and have little influence on each other. Second, we aim to emphasize the decision-making element of research. Decisions need to be made to determine those actions necessary to proceed from stage to stage. Each decision will be influenced by previous decisions and will, in turn, place constraints on later decisions in the process. An in-depth discussion of the nature and extent of some of these constraints will be reserved for Chapters 2 and 3.

2 From theory to hypotheses

Figure 1.1 represents an illustration of the first three stages of the research process and the actions required to progress from one stage to the next. It also shows that research strategy can mediate between theoretical perspective and the research stages. The theory and problem stages of the research process are closely interrelated and, unlike other stages, their order in the process is interchangeable. The reason for this is that not all research problems are selected on solely theoretical grounds. The problem selected may be one of practical importance or simply a reflection of the researcher's area of interest.

The reasons why we carry out research in psychology can thus be

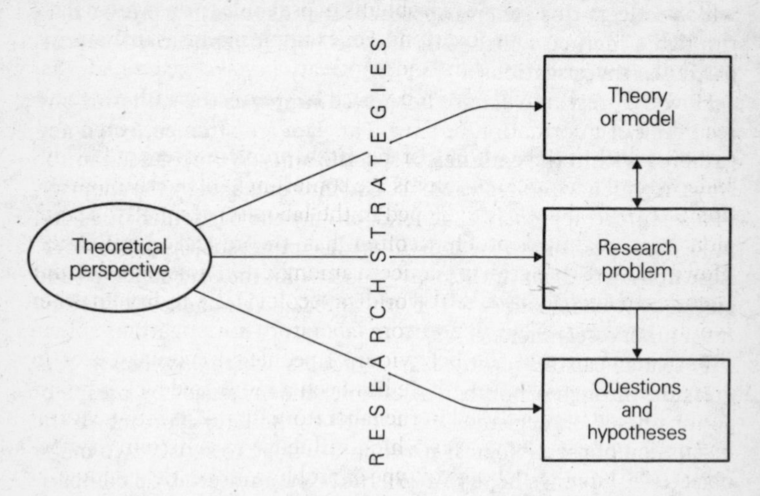

Figure 1.1 The research process: from theory to hypotheses

approached by addressing ourselves to the sort of problems that psychologists investigate. These can loosely be divided into those of a theoretical nature and those of a more practical nature. An example of a *theoretical research problem* might be 'How do we recognize two-dimensional representations of the three-dimensional objects found in the real world?' (e.g. television). Whereas a more practical problem might be 'Is there a relationship between violence on television and violence in society at large?' Indeed, the impetus for much of the practical research comes from the existence of 'real' problems as defined by our society, and sometimes reflected in government policy. Often policy decisions are made on the basis of commonsense, without the support of sound research findings (e.g. the introduction of microcomputers as teaching aids in schools). As a result, there are many psychologists who are interested in the practical problems subsequently generated by this sort of policy decision.

Practical research problems which are not necessarily policy-orientated may be problems arising from practices related to industry (e.g. fitting the right man to the right job), engineering (e.g. ergonomic designs of aeroplane cockpits), education (e.g. assessment of children's special educational needs) and health (e.g. the treatment of mental illness) – to list but a few.

Research into theoretical problems in psychology does not necessarily have practical implications. For example, theories of memory, reading, group interaction and motivation have generated vast amounts of theoretical research which has had only limited practical value.

Naturally, the two types of research problem are not always independent. Practical problems are sometimes solved through the application of theories developed in the laboratory; equally, a solution to a practical problem often has theoretical significance. However, there is a growing concern amongst psychologists that our theories frequently lack 'real world' or 'ecological' validity. In other words, are theories developed from laboratory investigations able to describe and predict the behaviour of people in everyday life? In response to this question, there has been a move to develop procedures similar to those used in the laboratory for use in more natural settings. Some of the factors which influence or constrain psychologists, both in their choice of type of problem and in the conduct of the research, are discussed more fully in Chapters 2 and 4.

Neither theoretical problems nor practical problems exist in isolation. Problem selection can be based on a theory that you wish to test; alternatively, it may be that a particular research problem of interest to you will offer indications as to which theory would be the most appropriate to apply. Regardless of the reasons for selecting a particular problem, the precise definition of your research problem is likely to have been determined by some theory, model or general theoretical perspective. For this reason, it is important to appreciate the way in which the different terms theory, model and theoretical perspective may be used.

Although the terms 'theory' and 'model' are often used interchangeably, some would argue that there is a distinction between them. A *theory* can be viewed as a set of interrelated propositions, each of which has been tested empirically. A proposition can be thought of as a relationship between two variables, usually in the form 'if *x* then *y*' – for example, 'if you are frustrated then you become aggressive'. The systematic nature of a theory allows us to predict consequences, logically derived from the interrelationships between propositions. If these predictions are supported by empirical observation, then we have not only tested the theory, but also provided an explanation of our findings.

A *model* frequently takes the form of an analogy or likeness to something. A model that represents a physical object is essentially descriptive (e.g. a model aircraft). However, a model can also be constructed to represent a framework of theoretical concepts (e.g. a computer model of human reasoning). This sort of model is more explanatory in the sense that it provides a conceptual arrangement which can generate testable hypotheses regarding the nature of the 'real' structure or process being investigated.

A *theoretical perspective* is a less formal conceptualization and represents a shared or common view of the world (or *metatheory*). One's theoretical perspective can have a pervasive influence on the research process as a whole and affects many of the individual stages. For example, two cognitive psychologists investigating memory may have two different theories to test, but both may share a common perspective or overall approach to a particular problem. Their common approach may be on a general level through the shared perception of man as an information processor or, on a more specific level, through the adoption of experimental procedures. This can be contrasted with the situation where two psychologists have different theoretical perspectives despite investigating the same problem. For instance, one may approach the definition of the problems of a maladjusted child from a humanistic perspective (e.g. via a psychodynamic approach to diagnosis) and another from a behaviourist perspective (e.g. via observation of the situational antecedents of the problem behaviour).

Certain theoretical perspectives are often associated with particular *research strategies*. There are three basic research strategies. First, there is the *descriptive* strategy, where the researcher sets out to describe rather than explain a psychological phenomenon. Good description is in most cases a prerequisite for viable explanation. It is difficult to explain something until you have described its essential properties. For example, if you were an industrial psychologist investigating poor productivity in a factory, you might well begin your study by describing the conditions of work, the nature of the work, the elements of the production process and so on. Having built up a good description of the situation, mostly through observation, you are in a better position to seek and demonstrate causes for (i.e. explain) the poor productivity.

This second type of research strategy is referred to as *explanatory*. This is a strategy which, through the adoption of certain data collection and data analysis techniques, attempts to explain a psychological phenomenon. An experimental investigation, which is designed to test predictions, is typically used by researchers who adopt this strategy. In our industrial example, you might speculate on the basis of your descriptive data that the poor productivity is a consequence of boredom. To test for this causal relationship you could design an experiment where half the workers are moved between jobs at regular intervals, whereas the other half remain on the same job for the whole time. If boredom was the critical variable then you would predict improved productivity for the first group. This research strategy would thus aim to explain *why* there was a low productivity rate.

The third and final type of research strategy is called *explorative*. This strategy is rather unusual since it can involve aspects of the other two strategies. However, the essence of the explorative strategy is that it does not involve the testing of precise predictions. A flexible approach is adopted which may involve elements of description and/or explanation, but which provides the researcher with tentative answers to his or her questions, as well as indications for the direction of future research. If, as an industrial psychologist, you are more interested in what factors make some of the factory workers more productive than others, then you may explore the relevance of a wide range of variables. Differences in performance on personality, manual dexterity or aptitude tests may be of importance. These would be descriptive variables (describing or establishing a profile of the productive worker). The importance of other variables may be established through experimentation (for example, changing the size of the team or the spacing of breaks). However, the importance of these variables would not necessarily have been predicted at the outset, and hence the study could only be considered an explorative one.

The tendency for certain research strategies to be traditionally associated with particular theoretical perspectives can create problems for the researcher. For instance, an anthropological approach to psychology (e.g. in cross-cultural psychology) is usually associated with a descriptive strategy. Equally, a cognitive approach to psychology normally adopts an explanatory strategy. However, your choice of research strategy should ideally be governed by the specific aims and considerations pertaining to your particular investigation and not simply by your discipline or that adopted by other workers in your area of research. The reason for this is that the research strategy you adopt will, as we shall see later, affect subsequent decisions in the research process.

One of the first decisions that will be influenced by your choice of research strategy is the way in which you formulate your *research questions* or *hypotheses*. Taking our previous productivity example, the precision of the research question may vary with the research strategy adopted. Investigators adopting a descriptive or explorative strategy may ask such questions as 'what are the relative productivity rates of the factory workers?' and 'what is the relationship between number of work breaks and levels of productivity?' respectively. On the other hand, an explanatory strategy should go beyond the simple question and formulate it in terms of a hypothesis. A hypothesis goes further because it relates two or more variables in an explicit way, and it takes the form of a testable statement. For example, a question about productivity rate could take the form of a hypothesis: 'Does an increase in the number of work breaks increase the productivity of the factory worker?'

Hypotheses are important for theory construction. Note how similar in structure a hypothesis is to the definition of a proposition given earlier in this chapter. Indeed a hypothesis could be called a propositional question. Hypotheses should be logically derived or *generated* from a theory or model, whereas other types of research questions need not be. It is because hypotheses can be derived from theories in a logical deductive way that they have an important part to play in research. A research hypothesis is testable. Therefore, the testing of hypotheses is one means of discovering the adequacy of a theory or model in its explanation of a problem. Consequently, if a hypothesis is not supported by empirical evidence, it may mean that the theory or model from which it was derived needs to be rejected or modified.

How are hypotheses generated from a theory? One potential problem with attempting to generate hypotheses arises from the abstract nature of theoretical concepts. It would be difficult to generate testable hypotheses without first defining your theoretical concepts in empirical terms, i.e., in terms of the observable, measurable properties of the variables under investigation. An *operational definition* is an attempt to define a theoretical concept by isolating empirical indicators. An example, from social psychology, of an operational definition could be that of 'class'. The definition of class may be in terms of people's income level. Those earning above £10,000 could be categorized as middle class, and those earning below that amount categorized as working class. Some researchers may consider such a quantitative measure unsatisfactory, and believe that class can be better defined by the sorts of job people do (e.g. white- or blue-collar work). To operationalize this more qualitative difference we may choose to define class on the basis of occupation. It is important to note that these two sorts of operational definition could require the application of quite different measurement systems or scales. This is an illustration of where, at an early stage, your decision about the operationalization of your concepts can limit your choice of measurement scale, and as we shall see in Chapter 3, your subsequent choice of statistical technique.

Furthermore, there is a danger that the operational definition can become divorced from the theory that generated it. For example, an acid could be defined simply as something which turns litmus paper red. The red colour of the litmus paper provides us with an empirical indication of the presence of an acid but does not represent a full description of the essential properties of acids. Similarly, some psychologists would argue that intelligence should be defined as 'what intelligence tests measure'. Definitions of this nature may be useful in the early stages of a developing science, but in general their value is strictly limited because they tend to be circular and self-prophesying. They should thus be treated with care.

An initial hypothesis, therefore, only states an expected relationship between variables as generated from a theory or a model. It does not tell you where these variables may be found or how they should be measured. In order to move from your hypothesis to your choice of data collection technique and subsequent research procedure you need to *operationalize* your hypotheses. For instance, an operational hypothesis in psychology could be 'if a factory worker is subjected to an increased level of noise at work, he will be more likely to make an error'. An operational hypothesis therefore provides indications as to how the concepts should be measured, and the population or target group from which you should select your subject sample.

3 From data collection to data analysis

Figure 1.2 outlines the particular aspects of the research process to be discussed in this section of the chapter. The research strategy that

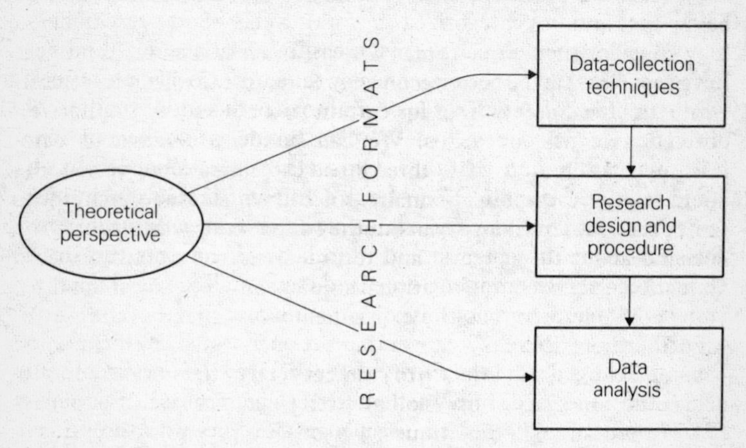

Figure 1.2 The research process: from data collection to data analysis

you adopt will not only affect the nature of the hypotheses or questions that you generate, but it will also influence your choice of research format. This in turn may limit the range of data collection techniques and forms of statistical analysis that are available to you.

Once you have decided on the research strategy most appropriate to your particular investigation (i.e., descriptive, explanatory or explorative) and have formulated your research questions or hypotheses accordingly, you will then need to decide on your general *research format* or combination of research formats. A research format can be defined as a particular type of investigative approach. There are two main formats characteristic of psychological research: the *experiment* and the *correlational study*.

An explanatory research strategy tends to rely heavily on the experimental research format, whereas correlational studies are mainly associated with descriptive and explorative strategies. For example, they are often used to describe people relative to one another (e.g. measures of extraversion) and explore the relationships between variables (e.g. the relationship between intelligence and impulsivity). Explanatory strategies cannot rely on a correlational format alone, because most correlational studies are retrospective. Consequently, although causality may be implied, it cannot be directly inferred. The essential differences between the two research formats, the factors influencing your choice of one or other of them and their effects on aspects of the research process will be discussed fully in Chapter 4.

The type of research format you choose will tend to limit your choice of *data-collection technique*. Some data-collection techniques are more appropriate for certain formats and by implication certain research strategies. We can divide the range of data-collection techniques into three broad categories: observation, the interview and the questionnaire. Of course, all three techniques essentially observe behaviour but they differ in the way in which the observations are mediated and recorded and the overall research situation employed. Correlational studies can adopt any of the three types of technique, but the experiment relies mainly on observational techniques.

Again there is a tendency to connect certain theoretical perspectives and branches of psychology with particular research formats and data-collection techniques. Personality psychologists tend to adopt a correlational format and commonly collect their data by means of interviews and questionnaires. Equally, cognitive psychologists frequently depend on controlled observation through their use of the experimental format. Naturally, there are good reasons for these associations, but equally there is sometimes a too rigid adherence to one particular format or technique. In the final analysis, the researcher should try to choose the strategy, format and technique that will provide the most useful information to help solve his or her research problem.

Do not assume from the above discussion that you should choose only one research format or data-collection technique. There is a growing belief that a combination of strategies, formats or techniques is often preferable to a reliance on any one method alone. Explorative strategies, in particular, may require more than one format or a variety of different techniques. In psychology, we use the term *multi-method* to refer to the adoption of a combined approach to research, and Chapters 2 and 12 discuss some of the recent developments in this area.

Your chosen data-collection technique provides guidance for the general method of data collection, but it does not provide any detailed directions regarding how you should actually carry out the research. At the *research design and procedure stage* you need to make decisions about your research design, your sampling procedure, your sample size, the precise details for 'running' subjects, the materials or apparatus required and many other questions relating to the actual collection of your data.

Yet again we find that such decisions cannot be made in isolation. The choices you make may well influence later decisions and are frequently affected by decisions already made. For instance, you may find that your choice of research design and sample size will limit your subsequent choice of data-analysis technique. In experimental research, you will certainly find that design and data analysis are two inseparable stages, because the statistical tests used in the analysis will often have explicit design constraints. Some of the influences on research design will be discussed in more detail in Chapters 3, 4 and 5.

Your data-collection technique will also provide little guidance as to how you should measure the variables under study. This will have been determined by the operationalization of your theoretical concepts which will also have influenced your choice of data-collection technique to some extent. Operationalization enables the relevant variables to be isolated and defined, and the process of *scaling* allows us to quantify them, thus enabling us to compare variables and to note changes within a particular variable. This quantification is achieved through the assignment of numerals to the variables under investigation, and to the various levels within each variable. The scaling of variables has to follow certain rules in order for you to create one of the four main scales of measurement: a nominal scale, an ordinal scale, an interval scale or a ratio scale. Chapter 3 provides a full description of these four scales and discusses the problems involved in choosing and creating the measurement scale most appropriate to the nature of your subject matter and data. At this stage, it is simply important for you to note that your choice of measurement scale in itself will be guided by all the previous stages and actions and will influence how you design your investigation, and collect and analyse your data.

Having collected your data, the next stage in the process is *data analysis*. However, there may be times when the data you have collected will not be directly amenable to analysis. Perhaps your data are not in numerical form, and thus may need to be *coded* first in order to facilitate analysis: e.g., data collected by the use of questionnaire and interview techniques normally require coding. Alternatively, you may wish to summarize or organize your data to provide a more manageable and meaningful representation (e.g. calculating means to summarize group performance). We use *descriptive statistics* to help us to describe, summarize and organize data.

With most research strategies you tend to collect data from relatively small groups or samples, with the intention of making inferences about a larger group or population. (A sample is a subset which should be representative of the larger population from which it was drawn – see Chapter 5 for further discussion.) *Inferential statistics* allow us to make inferences about populations on the basis of our samples. All three research strategies can use inferential statistics, but they do so in different ways. For example, we could make an estimate of a population mean based on our sample mean (i.e. describing the population characteristics). On the other hand, by adopting an explanatory strategy, you are more likely to use inferential statistics to test differences between means on the basis of predictions (i.e. to explain those differences). The explorative strategy may use inferential statistics in either or both of these ways (i.e. to describe and/or explain). The nature and uses of both types of statistics are explained more fully in Chapter 5.

In psychological research, the data obtained from an investigation are frequently analysed through the *application of a statistical test*. Your choice of test will be influenced by your research format. Correlational studies tend to use tests that measure the relationship between two or more variables (e.g. Spearman's rho) or the strength of association (e.g. Goodman–Kruskal's lambda). As already noted, the experiment is more concerned with tests for differences between groups (or populations, e.g. Mann–Whitney *U*) or means (or population means, e.g. *t*-test). However, many of the statistical tests at our disposal are not suitable for analysing and drawing inferences from data obtained from the study of a single case. Therefore, you may find that the desire to use a certain statistical test will limit some of your other research decisions.

The sorts of constraint that your choice of statistical test may place on the whole of the research process are so important that we have devoted much of Chapters 3 and 5 to their discussion.

4 From interpretation back to theory

Figure 1.3 provides an illustration of the major points relating to the concluding stages of the research process. The interpretation of the

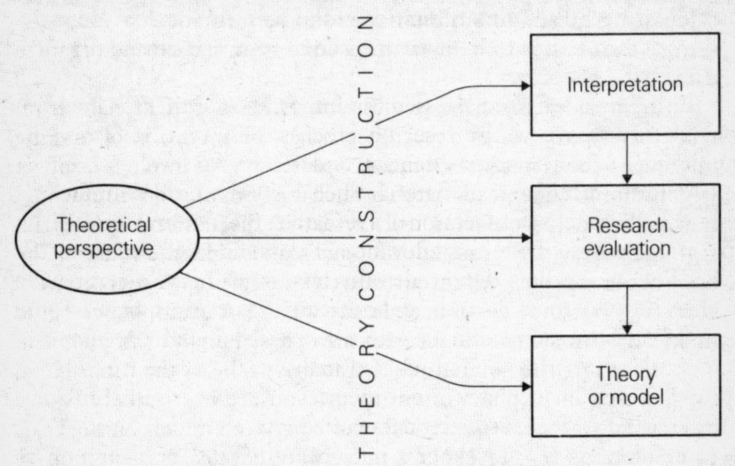

Figure 1.3 The research process: from interpretation back to theory

results of your analysis is guided and made sense of in the light of previous research findings. It is the researcher who interprets the data. This may seem obvious, but many students somehow believe that the significance level of an inferential statistic represents an interpretation in itself. Statistical tests aid interpretation, by determining any statistical significant result and hence indicating their importance. However, the theoretical significance of the findings must be determined by the researcher.

Although an evaluation of the importance of your results is usually aided by the application of a statistical test, it is not an essential component in research. Many Skinnerian behaviourists would argue that descriptive statistics alone are adequate, given that there is a consensus by the community that the effects are genuine. Statistical tests simply take the process one step further, by providing an assessment of the probability that your findings were produced by chance as reflected in the significance level.

Interpretation is thus a stage in the research process which is dependent on both theoretical and statistical guidelines as aids. Data are interpreted in the light of what is both theoretically and statistically significant. Therefore, by making interpretations with the help of statistics (descriptive and/or inferential), we are commenting on the adequacy of our models or theories. Again, as with many of the stages in the research process, your theoretical perspective can influence your interpretation of the data. As we shall see in Chapter 2, this influence can result in unprofessional practices on the part of the researcher.

In this chapter we have stressed the importance of making decisions at each stage of the research process. However, it is also important to have a stage at which you undertake your own *research evaluation* on a number of criteria. Such a stage would be represented by the 'discussion' section of a research report. It involves not only a discussion about the relevance of your findings to the research field as a whole, but it also invites you to make a *critical appraisal* of your investigation, and consequently look for improvements in your research methods. After all, only through the development of sharp and precise methods and analytical tools can we truly test and probe the adequacy of our theories. A negative result may simply reflect an inadequate data-collection technique, an inappropriate form of analysis or a poor operationalization of your theoretical concepts. It may also indicate an illogically generated hypothesis which cannot, with hindsight, be seen as a proper test of the particular theory or model.

Clearly, research is evaluated not only by the investigator involved in carrying out the particular project, but also by other psychologists, particularly those who are interested in the same problem area. The findings from studies given high regard by the scientific community are likely to become widely accepted. Subsequently, they may prompt other researchers to follow up or replicate the investigation, or to incorporate the findings into the theories they may be developing. However, research is rarely evaluated from an entirely unbiased viewpoint, and some of the grounds on which the evaluation may be based will be explored in Chapter 2.

The criteria for research evaluation are perhaps best illustrated through the consideration of certain probing questions. These questions concern decisions made at previous stages in the research process. As already noted, inappropriate decisions may have been made at the theory or model stage. At the data-collection stage, it could be questioned whether another data-collection technique or a more or less structured form of your existing technique might not have yielded clearer results. Equally, at the research design and procedure stage there is a need to consider any confounding (uncontrolled-for) variables which might offer an alternative explanation of your findings. Also, it could be questioned whether your procedures were really suitable, or whether they could have contributed a high degree of error to your results. Other important questions might be 'was an appropriate measurement scale chosen?', and 'was the form of statistical analysis the most suitable for the type of data obtained?' These and related issues are dealt with in Chapters 2, 3 and 5.

Of course, this list of questions is not exhaustive. The important point to grasp is that there is a need for you to stop, and take a critical look at what you have achieved. The importance of that achievement should be evaluated by you, as it will be by others, in the light of all the problems associated with psychological research. It is only when you feel satisfied with your answers to these critical questions that you should ask yourself the general question 'can my findings be incorporated into a theoretical system?' Your answer to this question will be based on considerations of whether your findings and/or your research setting and chosen population are generalizable. The making of *empirical generalizations* is an essential part of *theory construction* and development, and thus enables you to complete the research process, or allow others to do this for you! Some of the problems of generalizability, as they relate to the two main research formats, will be outlined in Chapter 4.

In order to demonstrate our holistic approach to the research process the various stages are now drawn together (see Figure 1.4). As can be seen, theoretical perspective is viewed as the central force in the process.

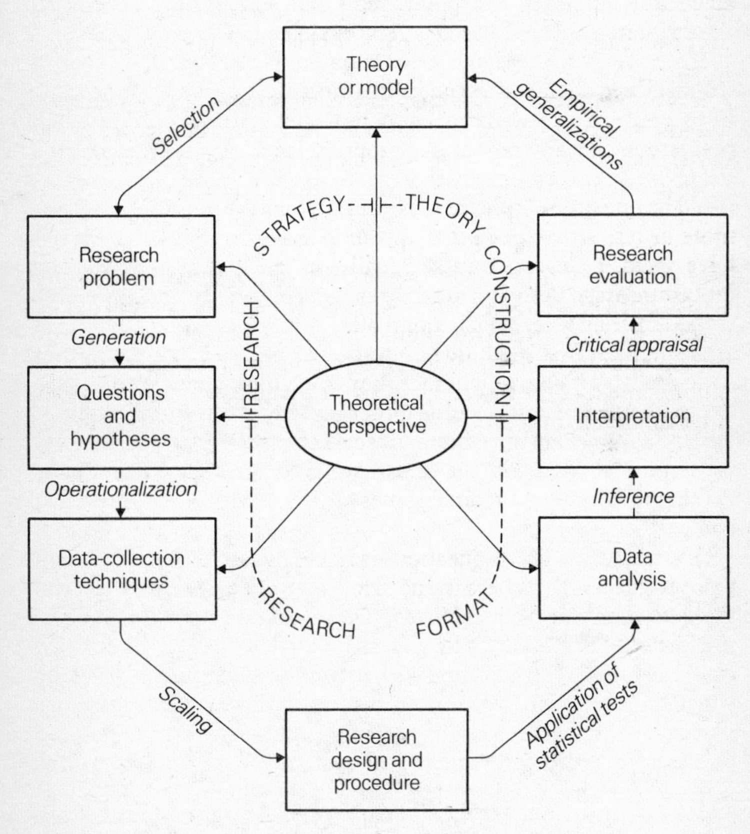

Figure 1.4 The research process: stages and actions

Summary

In this chapter research was presented as a process involving various stages, ranging from theory through data collection and analysis to interpretation and generalization. Emphasis was placed on the interdependence of many of the stages by stressing the importance of the decision-making element of research, in terms of the constraints a particular decision may place on other activities in the process. The general aims of this first chapter were not only that it should introduce you to many of the concepts developed in later chapters, but that by adopting a holistic view of psychological research it should also encourage you to perceive the various aspects of the process in the context of the overall objectives of your particular investigation.

It is also hoped that this chapter has increased your awareness of the important aspects of research, and the various influences on the decisions you will have to make when undertaking a psychological investigation. This increased awareness should help you become not only a better all-round researcher, but also a more critical consumer of published research findings. We tend to be inundated with 'statistics' from many sources. We are continually being presented with results suggesting, for example, that we should not eat certain foods or that we should inoculate our children against certain viruses. Unfortunately, such findings are often presented out of the context of the research process as a whole, which makes their full evaluation difficult.

2 Some Constraints on Research: Social and Political

The research process outlined in Chapter 1 does not exist in a vacuum. Research exists within a broader social context and the influence of this context has its greatest impact on research in the behavioural sciences, because of the nature of its subject matter. As a consequence, factors exist which may influence or constrain aspects of the research process. Some of these constraints arise from political or moral considerations, but more commonly they result simply from the problem of using people to study people.

In keeping with Chapter 1, we will consider how these social and political influences may constrain some of your activities at certain stages in the research process. We will also discuss ways of controlling for, or overcoming, the bias which may result from their influence. Some theoretical and statistical constraints which may also affect aspects of the research will be discussed separately in Chapter 3.

Problem Selection

The first activity in the research process to be influenced by social factors will be your selection of a research problem. As we have seen in Chapter 1, a research problem may be investigated from a non-theoretical standpoint. You may be interested in a problem because of its social or practical importance, or for no other reason than to satisfy your own curiosity. Other research may arise out of existing theories or areas which have become topical or popular and which have consequently created a 'bandwagon' effect. This popularity is often associated with the attraction of grant funds, as well as an increased likelihood of work in the area being accepted for publication or viewed favourably as a research proposal.

Business corporations and governments (often through their research agencies or their armed forces) allocate money for research. However, applications for funding have to be approved and favoured problem areas tend to be overtly stipulated. Research agencies may thus influence problem selection by approving projects of a certain kind, for example only those which they consider to be more 'applied' or to have potential practical payoffs. The larger pharmaceutical companies finance a great deal of research into the psychological effects of drugs, with the aim of marketing safe and acceptable cures. Likewise the armed forces are interested in research that will help improve the man–machine interface so that personnel can interact with their technology more efficiently. Government departments concerned with social welfare may give money for research into the problems of inner-city dwellers, particularly with reference to ethnic minorities or poverty.

Systematic research can often be expensive in time and personnel, and may therefore require financing beyond the researcher's own resources. In accepting external funding you should realize that you may lose some freedom in the choice of research problem. Although it may appear that these political constraints have little effect on the undergraduate's selection of a research problem, the research bias of the department will often limit the choice. This bias will be in terms of areas of interest and expertise and of availability of equipment and the subsequent effects on both the quality and quantity of research supervision. This bias will, to a certain extent, be a consequence of external influences.

Barnes (1979) argues that many inquiries into human behaviour can be seen as a 'process of interaction and negotiation between scientist, sponsor, gatekeeper and citizens'. With the term 'gatekeeper', Barnes is referring to individuals within groups or institutions who control the access of the researcher to their members or employees. For example, if you wish to question members of a particular trade union, you would need permission from a union official before you could approach them. One consequence of this interaction and negotiation between the different parties may result in you basing some of your research decisions primarily on what is acceptable to society at large and to the particular individuals concerned.

Societal values, as well as public interests and demands, can affect both your choice of problem and its definition. Certain types of investigation may be encouraged (e.g. the possible effects of high blood lead levels on children's cognitive functioning), whilst others may be deterred (e.g. genetic engineering) because of the ethical considerations involved. In practical areas in particular, social needs are likely to determine the priority areas for research interest and consequently may prompt you to select one problem rather than another.

Your personal attitudes and ideological beliefs may also influence your choice of problem area. These may often be shared by fellow researchers. For example, many psychologists would not undertake research which aims to isolate genetic differences in behaviour across racial groups, or which involves certain kinds of work with animals. Similarly, a group of psychologists may share a common perspective on behaviour and the way in which it should be studied. If you subscribe to a particular view it may limit your choice of problem and, more importantly, shape its definition. Finally, your personal interests, needs and career aspirations will undoubtedly influence problem selection.

Forcese and Richer (1973) refer to the influence of these sorts of social and political factors on problem selection as *prescientific*, because they impose constraints on the research process prior to the actual conduct of the research. They also stress that these influences do not affect the scientific nature of the research and are thus *extra-scientific criteria*. However, Forcese and Richer may be underemphasizing the role and effects of such constraints in the growth and development of science in general and in the behavioural sciences in particular.

Research Design and Procedure

Extra-scientific factors of a social and political nature can also have an important influence on the way in which you design and conduct your research project, particularly in terms of your choice of procedure and apparatus.

Those institutions and corporations that can affect your choice of problem can also influence how you carry out the research. They may restrict you in terms of the tools and procedures available. Research equipment such as computers or video equipment can be expensive, and therefore is likely to need to be approved and provided by your financial sponsors.

Equally, the very existence of equipment, particularly expensive equipment, can itself constrain the researcher. Problems become defined and operationalized in the context of existing and available research apparatus or test material. For example, there is a danger in thinking of traditional learning theories in terms of their experimental procedures or paradigms. In cases like these, theory can become paradigm-specific or problem-bound. We tend to think of Skinner's theories in terms of his (Skinner) box, or Thorndike's in terms of his (puzzle) box. The danger here is that your choice of equipment may predispose you to a particular theoretical bias. If you use a puzzle box it is very difficult to demonstrate anything else but trial-and-error (or non-insightful) learning in problem-solving. You should also be wary of the 'enabling' aspects of equipment. The existence of equipment often enables you readily to collect a certain type of data which is easily reproducible, and this may influence or bias your research. However, many of us are willing to work within the constraints of existing equipment in order to save the time and expense of developing valid and reliable new equipment ourselves.

By the very fact that in psychology we are usually manipulating situations involving human beings, we are restricted in the range of designs and procedures we are able, or willing, to use in order to investigate our ideas. There are situations where public opinion or interest may inhibit or prohibit certain investigative procedures. This may be because such procedures are considered to be immoral, if not illegal. To deprive children of physical sensation, in order to unravel whether perceptual processes are learnt or innate, would be both immoral and illegal. Many people consider experimentation on animals to be of doubtful morality although in most cases it is quite legal (with the appropriate Home Office licence). The public concern over the temporary blinding of cats in Blakemore's perceptual experiments (1973) was probably of sufficient intensity to make other researchers question the use of that particular procedure in their future experiments. Such situations do present psychologists with a particular problem. If a procedure becomes unusable at a later date because of ethical considerations it makes direct replication impossible. The possibility of replication is essential to a scientific approach. The experiment by Milgram (1963) on obedience and conformity provides a good example of this sort of situation. First, misleading or deceiving subjects into thinking that they are giving people severe electric shocks is now considered by most to be unethical. Second, the experimental details have received wide publicity. For either or both of these reasons this, and similar experiments, would be difficult to replicate.

The above discussion has focused on the extra-scientific constraints which may be imposed on your research, either by society in general or by social institutions. There also exist social influences of a more *personal* or *interpersonal* nature. These can be viewed as having an intrinsic or *intra-scientific* effect on the conduct of your research, because they may reduce the objectivity of your observations or the scientific rigour of your investigative procedures. Some such influences are the result of factors directly attributable to you as the researcher (e.g. experimenter bias or observer effects). Other elements of bias may be due to the particular characteristics of your subjects or research setting (e.g. experimental environment or interview situation). More importantly researcher, subject and situational factors can interact to produce yet further bias. It is recognized that individual subjects are likely to react to both the researcher and the research setting. Psychology, perhaps more than any other science, suffers from this problem of *reactivity*: subject reactions may produce unwanted effects which could mislead you or confound your research findings.

Let us consider the possible problem of subject bias first. It is normal practice for researchers to ask for volunteers to be subjects in their investigation. This raises the question of why some people volunteer whilst others do not. Does the volunteer group have any qualities or characteristics which set them apart from the non-volunteers? Rosenthal and Rosnow (1975) have presented convincing evidence to suggest that the volunteer is different from the non-volunteer subject. Volunteers are usually more intelligent, more sociable, better educated and from a higher social class. In other words, 'volunteer' samples may not be representative of the population at large because strictly speaking they were not constituted by a random selection from the population (see Chapter 5). The problem is increased by selecting most of our subjects from a limited population – for example, college students. Claxton (1980), however, would defend the use of college students in studies of cognitive functioning. He argues that if you are interested in footballing skills you would study footballers, so for investigations into cognition it is acceptable to study students. In certain situations such restricted selection can be justified, as long as the researcher randomly samples and limits his or her inferences and generalizations to that population. The nature of your chosen population should thus always be clearly stated.

A further potential problem for unbiased sampling is that subjects are not always *naive*. They may have an understanding of the concepts and procedures underlying the investigation, or they may simply have developed a level of 'test sophistication' from previous experience. The habit of repeatedly drawing samples from the same population of subjects certainly does not help to ensure subject naivety.

Besides the more specific subject characteristics outlined above, other personal attributes of the subject may influence the research findings. These include *biosocial attributes* (e.g. race, age, sex and class) and *psychological* or *psychosocial attributes* (e.g. intelligence, need for approval, anxiety level, personality and previous experience). For instance, males and females may respond differently to certain tasks, or highly anxious subjects may focus more on themselves than on the task at hand. As already mentioned, a more general problem arises from the fact that humans react with the people and situations they become involved with. This reactivity has cognitive, social and emotional components. Subjects are thinking, feeling and socially responsive beings, and this fact both restricts our procedures and confounds our results. Each subject approaches a particular research situation with expectancies based on past experiences. He or she will also have perceptual and mental 'sets' which influence subsequent responses, and the natural tendency to attempt to reason out the object of the investigation and continually test out hypotheses by utilizing any available feedback.

Subjects are attentive to social cues and bring their own learned social responses and individual social needs to the situation. Their emotional reactivity can influence performance, and an individual's emotional state can affect his or her concentration, motivation and attitude to the procedures. This may consequently result in unrepresentative responses.

Turning our attention to the influence of you, the researcher, similar personal attributes can again be of importance. All researchers, because of their previous experiences, also bring their own preconceptions and expectations to each investigation. These will then have some influence on how they perceive events, and therefore which behaviours are observed, recorded and responded to. For example, on a cognitive level, previous research experience may affect both the choice of design and the carrying out of the procedure. On a more emotional level, an anxious researcher may produce different results from those of a less anxious one. In addition, there is a tendency for an observer to see what he or she wants or expects to see. Rosenthal (1976) cites many interesting examples of this effect across a whole range of disciplines, including a study by Cordaro and Ison (1963), who asked observers to note the number of head nods and body contractions made by some flatworms (planaria). They manipulated *observer expectancies* by suggesting to separate groups of observers that either high or low rates of these behaviours were likely. They found that the observers fulfilled their own expectancies, although there was no real difference in the behaviour of the worms. Here the observers are unlikely to have influenced their observations intentionally. Unfortunately observer effects can also result from intentional falsification, usually at the data collection or interpretation stage, and these will be discussed in more detail in the next section.

In the above example, the observers had little direct effect on the behaviour of the subjects (the flatworms). Further research bias is introduced when certain observation and interview techniques are used, however, because the individual characteristics of the subject and the researcher are allowed to *interact*. The problem of the subject reacting with the observer is one which mainly, but not exclusively, concerns the social sciences. Rosenthal's (1976) account of the area is strongly recommended. What is important to note is that there are certain attributes of the researcher, his subjects and the research setting which may make demands on a particular subject to behave in a certain way. Orne (1962) has described these variables as *demand characteristics*. The interaction of the subject and researcher's biosocial attributes are the most easily understood. For example, there is evidence that an attractive female experimenter produces more learning from her male subjects than a 'husky' male experimenter (Binder, McConnell and Sjoholm, 1957). In the area of race and class, Labov (1970) has highlighted the problems associated with using a white, middle-class experimenter to assess the linguistic performance and competence of black ghetto children.

The effects of psychosocial attributes are less obvious but no less important. Of particular importance is the notion of 'need for approval'. The researcher and the subjects may feel the need to respond in a way that will gain each other's approval. Researchers may also want to produce a successful piece of research so that they can gain approval and recognition from their colleagues. This may be translated into implicit demands on their subjects to confirm the research aims or hypotheses. In the case of the subject, a need for approval takes the form of *evaluation apprehension* (Rosenberg, 1969). Subjects may feel, justly or unjustly, that they are being assessed or evaluated. Therefore they may wish to present themselves in the best possible light, instead of responding in the most spontaneous and genuine way. More indirectly, they may simply wish to gain the approval of the investigator by performing to his or her expectations. The problem is that not only do researchers communicate or 'leak' their expectancies, but subjects, even in the most rigorous conditions, seem intent on confirming those expectancies!

Rosenthal (1976) discusses in some detail examples of how *situational factors* can create a reactivity problem. In these cases the reaction is not to the researcher *per se* but to the research setting and procedures employed. Naturally the researcher is part of the research setting, and so experienced researchers may obtain results where more naive researchers fail. However, there are also particular characteristics relating to the research situation which can influence behaviour. These include factors like the physical features of the immediate environment, the equipment or materials used in the investigation, individual *versus* group participation, and personal *versus* impersonal contact with the researcher. Let us elaborate further on these situational aspects. The size, temperature and contents of the test or interview room may well affect the quality and quantity of subjects' responses. The nature of the research procedure can also affect performance, in that subjects may respond differently in a group setting from the way they would do on an individual basis. They may also react differently to test items being presented visually as opposed to orally, or to a computer presentation instead of a personal one. Some of these problems will be discussed later, in relation to possible ways of controlling for situational bias.

A further category of possible influences which needs to be taken into consideration contains those factors which arise out of the *person–situation interaction*. One example of this type of factor might be the effects of a very introverted subject taking part in a *group* experiment or interview. The personality of the subject would be likely to interact with the particular demand characteristics of the social situation involved, so that he or she may distort the results by reacting differently from the rest of the group.

Another, more subtle, sort of person–situation effect relates to the interaction between the biological state of the subjects at the time at which the investigation takes place, and the nature and demands of the task presented to them. There is increasing evidence to suggest that a woman's performance on certain tasks varies according to her position in the menstrual cycle. Equally, an individual's performance on a particular task, for instance a test of memory span, may change over a day or a month. In most cases these effects, resulting from cyclical changes in a person's biological rhythms, are small and their significance is only realized on particularly sensitive tests and procedures (Folkard, 1982). However, they can also have a more pervasive effect on people's moods, which may in turn affect their perception of task requirements and hence performance (Isen, 1970).

Controlling for the bias which may arise from problems of reactivity can be approached in two ways. First, we can tighten up or introduce new controls, tools and procedures which may limit or confine the extent of the problem. Second, we can adopt new data-collection techniques which, by design, try to prevent or reduce the interaction of the observer and the observed. Such techniques or methods are not surprisingly called *unobtrusive*.

In the first case, the use of a highly *structured* data-collection technique and standardized procedures can overcome *some* of the reactivity problems. For instance, a structured interview will allow less scope for uncontrolled interaction between the interviewer and the interviewee. Equally, strictly defined recording techniques should reduce the likelihood that subjective or selective observations of behaviour will be made. The degree of structure can also refer to the general setting in which the data are collected. Highly structured settings (e.g. the laboratory) are obviously rather artificial, but in these situations the researcher has greater control over events and can therefore maintain more *standard conditions* – i.e. the same situation is presented to each subject.

There now exist a number of procedures which have been specifically developed to help control and reduce subject reactivity. One classic example is the use of a *placebo*. Typically, this procedure is adopted when a researcher is testing the effectiveness of a new drug.

If the problem of reactivity did not exist then the researcher could give the drug to one group (the experimental group) and compare them with a group which did not receive the drug (the control group). However, the very act of giving the drug, perhaps in the form of taking a pill or having an injection, may well produce an effect that is not directly attributable to the actions of the drug itself. To overcome this problem, the researcher needs to be sure that both groups undergo identical procedures, which would include giving one group the 'real' drug and the other a 'dummy' drug or placebo. The important aspect of this procedure is that subjects are not aware whether they have or have not received the treatment. An even tighter control would be to make sure that, in addition, the researcher does not know which group received the drug and which the placebo, thus also reducing observer effects. This procedure is called the *double-blind* control. Unfortunately, a researcher with knowledge of an area might receive clues from the subject's behaviour, perhaps due to side-effects or simply from knowing what reaction to expect. Therefore, quite often, totally naive research assistants are asked to administer the drug and record its effects.

Although the above control procedures have their roots in the medical sciences, they do also have uses within experimental psychology. The essence of the procedure is to ensure that all groups receive identical, or near-identical, treatment in all aspects except the experimental manipulation. Under such conditions the researcher can feel more confident that any obtained result is not the product of differences in procedure between groups, but is a genuine and reliable treatment effect.

Essentially, what the investigator often has to do in these 'blind' conditions is to deceive some of the subjects or research assistants. Ethics normally require the researcher not only to debrief each subject at the end of the investigation, but also to brief the subjects beforehand with as much truthful information as possible, without actively inducing reactivity. This can be difficult to achieve. Resnick and Schwartz (1973), in a verbal conditioning experiment, forewarned their subjects about the nature of the experiment as required by professional ethics. They found, however, a 'boomerang' effect in conditioning rates which was quite contrary to the existing laws of verbal learning. For this and similar reasons, there may be experiments where the researcher actively and overtly sets out to deceive a subject by disguising the true purpose of the research. You may, for example, be interested in how people behave in emergencies. However, to reduce the problem of reactivity it might be necessary to hide your intentions from your subjects by telling them that you are interested in how small groups of people go about solving problems. You could then arrange for a fire-bell to ring in the middle of the session, or perhaps for smoke to be blown into the room. It is at this point that your 'real' experiment begins, as you observe their reactions to your smoke or bell (Latané and Darley, 1970).

Many researchers consider these disguised experiments unethical. One possible way of overcoming the ethical problems involved with deception is to attempt to alter the subject's role. In these situations, the subjects are told what the manipulation will be and what is expected of them. The subject is then asked to play the role of being a subject. There are a number of problems with this approach. First, to be successful as an experimental procedure the subjects are still only partially informed. Second, as researchers like Miller (1972) and others have pointed out, people find it difficult to anticipate their potential or future actions or behaviours. Finally, even in role-playing, the participants are still taking part in an experiment, and hence their role as a subject may take precedence over, or conflict with, the other experimenter-determined roles. Perhaps the strength of the *role-playing procedure* lies more in preparing the experimenter, through the active participation and co-operation of the subject, for possible reactivity effects which can then be controlled for in the main experiment. In other words, it could be a useful pilot procedure.

The control procedures discussed above address themselves mainly to a reduction in the cognitive aspects of reactivity. An attempt to control for social reactivity would be to *distance* yourself from your subjects in some way. There are several techniques and procedures within psychology which may help you to achieve this aim. You could, for example, observe the behaviour of subjects indirectly. In order to do this, a *one-way mirror* or *video equipment* could be used. However, assuming that you have briefed your subjects, to satisfy the ethical code, you may still find some reaction to the presence of cameras or mirrors. This reaction can be reduced by allowing time for the subjects to adapt to the presence of the equipment. Data collected from the first few sessions or test trials could then be disregarded in the data analysis stage. Indeed, such a procedure may be wise in all psychological research, as subjects may well be anxious or apprehensive at the start of a session, where the effects of social and emotional reactivity are generally more apparent. Another traditional approach to this problem is to *counter-balance* treatments. By varying the order of treatments, so that no one treatment is predominantly undertaken in any one position, you can help to ensure that extraneous effects like anxiety (at the beginning of a session) or fatigue (at the end of a session) do not contribute a constant error to your design. By counter-balancing, a potential constant error can be transformed into a random error (Robson, 1983, pp. 34–6). Although random errors are infinitely preferable to constant errors, it is probably sensible to attempt to reduce both by adopting a settling-in period *and* a counter-balanced technique when designing your investigations.

Distancing yourself from the subject may also be achieved through a change in the data-collection technique. *Questionnaires*, for example, do not usually require the presence of the researcher while they are being completed by the subject. In terms of reducing social reactivity, the use of a questionnaire may seem preferable to an interviewing technique. However, the use of questionnaires introduces problems of its own. One of these is the difficulty in assessing levels of comprehension and motivation. Although the adoption of a questionnaire data-collection technique overcomes many of the problems resulting from subject reactivity, cognitive and emotional reactions to the way in which items are worded and presented may still influence subjects' responses.

Another possible way of overcoming reactivity problems, especially reactions to the research setting, would be to adopt an alternative method of investigation, e.g. a *field study*. By 'field', we mean the subject's natural environment rather than an artificial setting; in our earlier 'emergency' example, you would attempt to observe people's behaviour in a real fire.

There are no easy solutions to the problem of reactivity. Some researchers have argued that the contribution of possible reactivity and other extraneous effects to the results can be assessed and reduced by the use of more than one method. Sociologists, for example (Denzin, 1970), have suggested that through the process they call triangulation, we can gain a better perspective on a research problem. In effect, they see the problem as if it were positioned in the centre of a triangle, which can then be viewed with differing perspectives from its three corners or angles.

In psychology, there has been an increase in the number of *multi-method* studies. This growth in the use of a multi-method approach should be welcomed, although psychologists should maybe focus more on the possible ways in which the various methods could be synthesized, rather than viewing them as essentially different though complementary approaches. For example, Mason et al. (1983) compared and contrasted the behavioural, Piagetian and information-processing approach towards the way in which learning and thinking develop. However, relatively little time was spent on synthesizing these approaches in order to provide a more unified perspective on the problem.

Data Analysis and Interpretation

The effects of social and political constraints on the data analysis and interpretation stages of research are also important. Scientists cannot avoid looking at their data from their own subjective viewpoint. What they find, and their interpretation of results, may thus depend on criteria which are not always entirely objective.

Particular problems arising from the social or cultural attributes of the researcher are highlighted in the field of anthropology. However, their importance for the behavioural sciences is increasing as our societies become culturally more pluralistic. The problem is essentially one of *ethnocentricity*. This problem relates to the interpretation of data concerning a particular culture or group of people in the light of one's own cultural experiences. A somewhat simple but clear illustration of this is to think of French and English anthropologists interpreting data about a people in a way that mirrors or represents a microcosm of their own social structures. Labov (1970), as we have already noted, demonstrated with some force that the problem of ethnocentricity exists, to a lesser extent, within one's own culture or society. He questioned studies which suggested that the language of the black American ghetto-dweller is crude and impoverished. Labov's own analysis demonstrated that the language elicited by a black researcher is quite different from that elicited by a white middle-class researcher. He also implied that ghetto language, when assessed in a less formal setting, is as rich as conventional English. Similar problems can be found in studies of animal behaviour. Here the problem is to avoid being *anthropomorphic*, that is, reading human qualities and attributes into your interpretation of the animal's behaviour.

Perhaps your greatest problem as a researcher is the pressure placed upon you to produce the 'goods', i.e. significant results. Academic job security or prospects are frequently determined on the basis of an individual's publication record. Likewise, as a psychology student you may feel that a project which reports statistically significant findings is more likely to be awarded a good mark than one from which no clear conclusions can be drawn. Research agencies, firms or governments which have spent money on a particular research project will be reluctant to renew or award new grants for associated research, unless the project has produced important or useful results. The journal publication system in particular puts pressures on individual researchers. The policy of publishing only statistically significant findings is quite understandable, but it may prevent a statistically non-significant, though theoretically important, finding from being published (e.g. a failure to replicate). These sorts of pressures can unfortunately result in a tendency for some researchers to 'fudge' or 'cook' results. The social and natural sciences both have their celebrated examples of discovered fudgers! Sir Cyril Burt, for example, appears to have obtained exactly the same correlation coefficients in more than one of his published twin studies – a very unlikely event indeed! Burt was already well known and established, and so pressures arising from the need for success and promotion would probably not explain his actions. Newton and Mendel are similar examples. The reasons in all these cases appeared to be associated with the overwhelming need to obtain data to fit an intuitively valid theory in a potentially important area. At the other extreme, Summerlin responded to work pressures by painting the skin of his mice in order to give the appearance of successful skin grafts. In Summerlin's case his academic future, in the absence of useful results, would have been comparatively bleak. There are numerous examples of individuals, great and small, who have succumbed to social or political pressure and intentionally fudged their data to fit their and/or their employer's expectations.

Other social pressures can have a more subtle influence and produce interpretational and statistical errors of a more unintentional kind. How do these influences become effective? Barber (1976) argues that there are five potential problem activities of which the researcher should be aware. First of these activities is *data trawling*, or the gathering of as rich data as possible. Problems can arise, especially when adopting an explanatory research strategy, if projects are too data-rich. The temptation is a natural one, because the more data you have the more chance you have of finding a significant result. However, this can also produce data which may be difficult to interpret and explain. As we saw in Chapter 1, there are situations where you may be aiming to explore rather than directly explain. In these circumstances, by means of an explorative strategy, you could collect a broad spectrum of data which may then provide the basis for a more detailed study at a later date. Research studies need to be carefully planned because the amount and complexity of the data required will depend on the nature of your research question.

The second potential problem activity relates from the temptation to change your research hypotheses to accommodate unexpected but significant findings. *Post hoc explanations* must always be speculative and therefore cannot be used to establish a causal relationship between two variables. If you do find an unexpected result, then use it in your discussion section as the basis for generating new hypotheses and planning new studies to test these hypotheses. Serendipity, or discovery by accident, is not to be under-rated but it must be presented and understood in its proper context. Barber also warns the researcher of the dangers of *post-mortem data analysis*. By this he means that you should avoid slicing or cutting data in previously unintended ways and subsequently carrying out new analyses. Again, this problem relates mainly to explanatory strategies, since any significant results arising from such activities should be used only as indicators for future research.

The pressure from journals to publish only statistically significant results can produce a tendency to repeat failed investigations until a significant result is produced. Statistically speaking, continual repetition must produce a significant result if only by chance! In such circumstances the final set of *data declared* will appear more valid and theoretically significant than it should do.

Barber's final caution for the researcher is that data tend to be checked for errors only if the analysis produces a contrary or non-significant result. *Data checking* is advisable, and indeed should become routine, in all data handling situations. In addition to the possibility of misrecording data, there is also the problem of inaccurate input of data into calculators or computers. Therefore, always check your data, regardless of how careful you think you have been. Computers will usually print out data for ease of checking. If you are computing your results by hand make sure you also check your calculations.

Finally, you must be cautious when considering the exclusion of troublesome subject scores. Naturally, good excuses are usually found for ridding yourself of seemingly perverse data! Sometimes there are good reasons for replacing subjects (e.g. a person with bad hearing doing an auditory based experiment, or a single 'wildly' discrepant score). However, the dropping, and subsequent replacement, of those subjects who distort your 'means' should not be undertaken lightly.

Research Evaluation

The same factors that influence or restrict your choice of research problem are likely to affect the way you report your results, how you evaluate their theoretical significance, the use to which you put them and the importance with which they are regarded by other people. Forcese and Richer (1973) refer to the factors influencing the evaluation and application of research findings as *post-scientific criteria*, because essentially they come into effect *after* the investigation has been completed.

The acceptability of your findings is determined partly by their level of statistical significance and partly by their theoretical or practical importance. Hence there may be some resistance to the widespread acceptance of certain findings because of their potentially undesirable political or social applications; equally, important findings with financial implications may not be acted upon or made public by the research sponsors. Alternatively, a finding which is only just statistically significant, or which is based on an unsound methodology, may be blown up out of all proportion because it supports a popular theoretical position or social policy.

Finally, it may be the case that your findings are both statistically and theoretically significant but nevertheless fail to gain the recognition of the scientific community. The scientific establishment will often stubbornly resist new theories to the extent that it takes a 'revolution' before they are widely accepted (Kuhn, 1970).

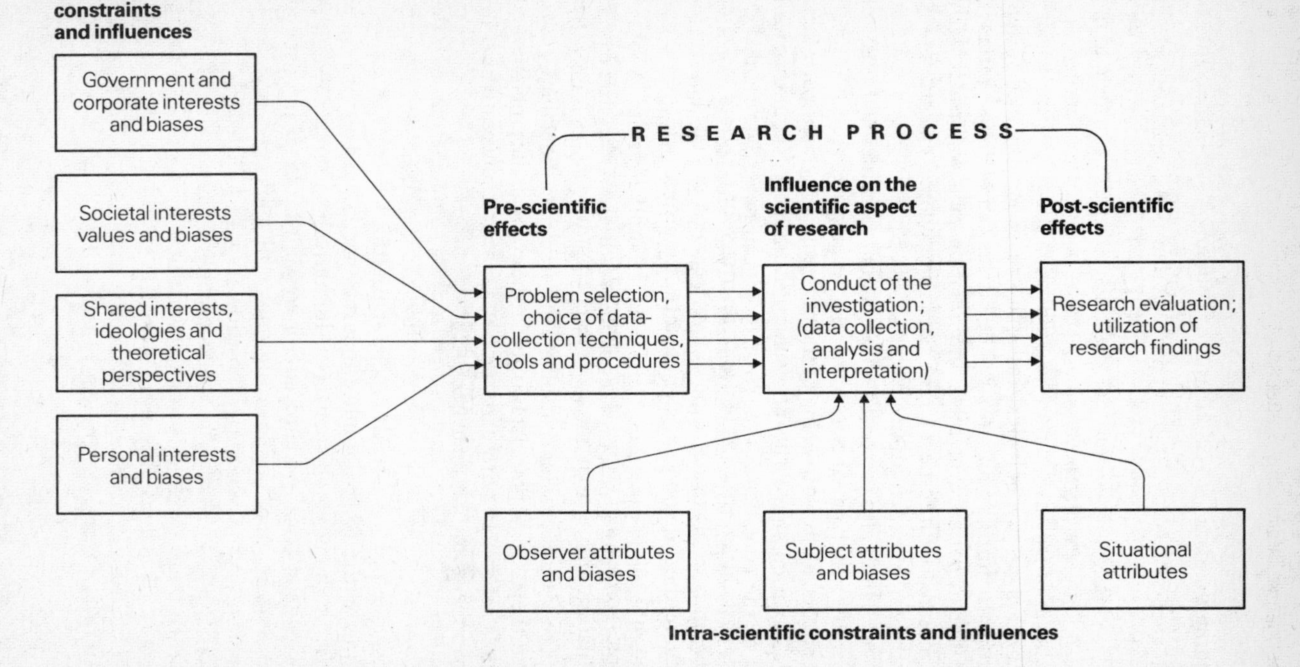

Figure 2.1 Social and political factors affecting the research process (an adaptation and extension of figures from Forcese and Richer, 1973)

Summary

This chapter described some of the social and political factors which may have an influence on your research. They were divided into those which could be considered extra-scientific or extrinsic to the research process (e.g. professional ethics) and those of a more intra-scientific or intrinsic nature (e.g. observer effects). These various influences on psychological research can be summarized in diagrammatic form by Figure 2.1. The factors can of course interact, and the importance of these interrelationships was particularly emphasized with respect to the need to recognize, and attempt to control for, intra-scientific effects.

The chapter also discussed a number of procedures and items of apparatus which have been specifically developed to reduce 'reactivity' effects, along with some of the problems associated with their use. Our discussion of the intra-scientific influences should not lead you to the conclusion that psychology can never be regarded as objective or scientific. One important way of safeguarding objectivity is through the use of replications. A finding is considered to be more objective if other members of the scientific community can repeat the same procedures, and produce similar results. Another reason for 'tight' procedures therefore is to make replication easier. An aim to eliminate all the social influences on research is unrealistic, if only because science itself is a social activity (Friedman, 1967). Nevertheless, psychologists do need to be aware of the problems and pitfalls associated with carrying out a research process which involves the study of human beings.

3 Further Constraints: Measurement and Statistics

In Chapter 2 we discussed some of the complications and restrictions arising from social and political influences on the research process. Further constraints arise from the effects of the measurement system, measurement strategies and statistical techniques that we adopt in psychological research. We will be focusing on these sorts of constraint in the present chapter.

Measurement Systems

Although measurement is a rather complex topic, some discussion on the matter is necessary, in order to help you understand why there may be measurement requirements associated with the various statistical tests found later in the book.

All measurement is based on mathematics, and in psychology it involves the application of abstract mathematical systems to the measurement of psychological variables. We can define measurement as *the assignment of numerals to objects, persons or events, according to rules*. In other words, this assignment is made through the application of a measurement system to our observations. However, measurement is rather a problematic issue for the behavioural sciences. It is more straightforward in the context of the natural sciences, because measurement theory was developed in direct response to many of their own particular measurement needs. The subject matter studied by these sciences is also generally more amenable to precise measurement. For precise and literal measurement to be achieved, the laws and definitions relating to the theories involved need to correspond directly to those of the mathematical system adopted (i.e., they should be *isomorphic*). Unfortunately, the nature of psychological theories presents difficulties in this respect.

They are usually too imprecise to indicate the appropriate means of measuring the relevant variables.

The discrepancy between the imprecise nature of our psychological theories and the precise nature of the measurement systems creates problems where the interpretation of findings is concerned. A particular set of observations may be assigned a range of numerical values, which do not reflect the *true* differences in magnitude of the property being measured. Depending on the nature of your research question and the sort of inferences you wish to make from your data, this may or may not be important. However, it can sometimes result in an inappropriate measurement system being applied to psychological data, and unjustified meaning attributed to the findings subsequently obtained.

In psychological research, problems arise even in the initial isolation and definition of those variables truly relevant to our research question. A particular difficulty associated with measurement in psychology is that the direct measurement of our variables is often impossible. Psychologists are frequently forced to rely on the measurement of another related variable, because many psychological concepts are not amenable to direct or literal measurement. We therefore often assign numerals to variables presumed, on *a priori* grounds, to be related to the concept under study. For example, we might measure how hungry an animal is on the basis of how many hours it has gone without food, or how much a person has learnt by the number of errors made on a test. We can measure only the external, observable behaviours assumed to reflect these 'hidden' (or *intervening*) variables.

The different measurement systems that we can apply to our observations result in different *levels* of measurement. These levels are represented by the four main kinds of scales: *nominal*, *ordinal*, *interval* and *ratio*.

1 Nominal scales

Nominal scales represent the simplest and weakest form of measurement. This scale is used to *classify*, name or label persons, objects or events. This qualitative assignment of observations to classes is not strictly speaking measurement. From our definition, it can be seen that the term 'measurement' usually refers to situations in which a numeral, rather than a label, is assigned to an observation. However, the subsequent element of counting involved has led to the accepted practice of describing the nominal level as a measurement scale. There are several properties associated with nominal scales which affect their use.

First, the assignment to classes or categories assumes that the classes are discrete or *mutually exclusive*: in other words, a particular observation cannot belong to more than one category. Thus the classification is not based on a question of degree, but on an 'all or nothing' judgement.

Nominal scales must also be *exhaustive*. Exhaustive means that the classification must take into account all sub-classes of objects or events. For example, imagine that you are interested in which political party a person will vote for in the next election, and that the parties are represented on a nominal scale. For the scale to be exhaustive, all political parties and not just the main parties need to be represented. Most researchers overcome this problem by adding a 'catch-all' category, usually in the form of 'any other'. An example of a nominal scale where this sort of category is not needed is the classification of people as male or female. Since you cannot (in theory) be both male and female the two categories are mutually exclusive, and as these categories cover all possibilities they are also exhaustive.

A common procedure, particularly when questionnaire or interview techniques are used, is to assign numerals to each category on the scale, in order to 'code' and thus simplify the data. For example, males may be represented by the numeral 1 and females by the numeral 2. The assignment of numerals to categories is thus purely arbitrary, so it does not make sense to use arithmetic operations *directly* on the coded data. The relation between members of a certain category is that of equivalence (=). This means that all members of the category are equivalent on the particular property being measured. Nominal scales allow the number of instances in each category to be counted easily and precisely. From this type of data, descriptive statistics such as frequency counts and the mode can be readily obtained. It is then possible to apply a statistical test to the frequency data, in order to answer questions about the distribution of cases amongst the various categories.

2 Ordinal scales

Ordinal scales represent the most primitive level of *numerical* measurement. Whereas nominal scales consist of discrete and 'absolute' classes, ordinal scales are concerned with the relative or *ranked* position of classes on the particular continuum being scaled.

Numerals are assigned to individual observations (e.g. to scores) according to their relative values so that they comprise a complete rank ordering of classes involved. However, although an ordinal scale provides information about relative positions or orders, it tells you nothing about absolute values. In a race, you could rank the competitors in the order they cross the winning line (first, second, third and so on), but the ranks will not provide any information about individual times or the time intervals between each competitor.

Ordinal scaling thus involves the relation 'greater than' (>) between classes, in addition to the relation of equivalence between members of the same class. Except for the median, the computation of descriptive statistics from ordinal-level data (e.g. the calculation of means and standard deviations) may have little meaning in terms of a psychological theory. However, the numerals representing the ranks of an ordinal scale may be summed, and other arithmetical operations applied to these totals. Statistical tests based on rank-order data may thus be applied to data measured on an ordinal scale; Chapter 10 mentions some examples.

3 Interval and ratio scales

Interval and ratio scales are regarded as truly *quantitative* scales and represent a higher level of measurement. This is because they not only subsume the rules that hold for ordinal scales, but also contain precise information on the magnitude of the interval between classes or between the points on the continuum being measured. On an interval scale the exact magnitude of the 'distance' between any two points on the scale is known. In other words, the unit of measurement is constant, despite the unit itself and the zero point being arbitrary.

Interval scales obviously have zero points, but these do not correspond to any 'real' or absolute zero. A consequence of this is that you cannot say that a particular value on an interval scale is twice (or any times) as large as any other. For example, the centigrade temperature scale has a zero point but its zero has no one-to-one correspondence in the real world. We cannot legitimately say that 40°C is twice as hot as 20°C. What we can say is that the interval between 20°C and 40°C is equal to the interval between 0°C and 20°C.

On the other hand, an income scale can have a true zero point, since there is in reality, as many of us know, an equivalent of zero income! It is therefore possible to say that Smith, with an income of £10,000, is earning twice as much as Jones, with an income of £5,000. Thus, the additional property of a *ratio scale* is that it does take the zero point as its origin. With a true zero point the ratio of any two points on the scale may be determined independently of the unit of measurement. For instance, if one building is twice as tall as another, it remains so irrespective of whether the two buildings are measured in feet or in metres.

For interval-level data, all the arithmetic operations may be applied to the intervals (or differences) between scores, and for data measured on a ratio scale the arithmetic operations may be also applied to the actual numerals assigned to the observations. Consequently, under certain conditions, not only can the useful descriptive statistics obtained from data measured on these scales include means and standard deviations, but also a much wider range of statistical tests can be meaningfully applied.

As can be seen from this discussion, the level of measurement achieved can limit the range of statistical tests available to you. It could be argued that arithmetical operations may be applied to virtually any numerical measurements, but we must consider whether the results obtained would always be meaningful from a psychological point of view. In Siegel's (1956) view, 'the scales used by behavioural scientists typically are at best no stronger than ordinal'. Siegel's claim refers us back to the various problems facing psychologists using measurement systems outlined earlier. As Hays (1973) points out, 'the road from objects to numbers may be easy, but the return trip from numbers to properties is not' (p. 90). Consequently, your choice of measurement scale should be considered carefully, in the light of your research objectives. If one of your research objectives is to make statements about the actual magnitudes of the properties you have measured, then you will need to be able to justify your use of a quantitative scale, such as an interval scale.

Many psychologists would now argue that by making certain assumptions, we *can* legitimately create an interval scale (Gaito, 1980). Some of these assumptions are described in the later section of this chapter on 'choice of statistical test'. Nevertheless, the important point to note is that the interpretation of statistical results lies with you, the researcher. The statistics cannot, by themselves, provide you with answers to your questions.

Measurement Strategy in Psychology

Psychologists have tended to adopt a particular strategy in the measurement of psychological variables, and this strategy involves the concept of *variability*. To remind you, variability is concerned with the spread of scores about a central point (usually the mean).

The importance of the concept of variability for psychology was initially demonstrated through the work of Sir Francis Galton in the nineteenth century. One of Galton's many interests lay in the measurement of intelligence. Galton assumed that the intelligence scores of a population would be normally distributed or, in contemporary jargon, have a distribution that obeys the Normal Law of Error. In the words of Johnston and Pennypacker (1980), Galton 'mapped the theoretical normal distribution onto a 14 step, equal interval scale and, in one grand gesture, invented not only the concept of intelligence, but a means of measuring it as well.' What this means is that the concept of intelligence has effectively been defined and measured in terms of variability (the normal distribution). Your level of intelligence therefore is defined both by your relative position on a distribution (assumed to be normal or bell-shaped) and on the amount of variability about the mean score. If the variability or spread about the mean IQ is small then an individual score which is considerably greater than the mean would be most uncommon. Thus the individual concerned could be defined as 'very intelligent' in comparison with the rest of the population. If, on the other hand, the variability or spread about the mean is increased, then that *same* IQ score may not be so infrequent. Consequently, the person would not be considered as being quite so 'unusually' intelligent. The statistical reasoning behind variability will be explored in Chapter 5, and its consequences for questionnaire design will be discussed in Chapter 8. The critical notion to grasp at present is that variability not only forms the basis for the measurement of many psychological variables, but is actually used in the definition of many of our concepts.

Although accounting for variability within and among natural phenomena is a fundamental aspect of any science, there are two problems associated with the application of this particular strategy to the measurement of psychological variables. First, some would argue that it is necessary to have standard and absolute units in order to obtain an accurate measure and description of the occurring variation. But, as in our example above, psychological measures are rarely standard or absolute. The units used to measure intelligence may change with our developing conception of the phenomenon. Units of weight or mass are both standard and absolute and, importantly – unlike our IQ example – the measurement *per se* does not form the basis for the definition of the phenomenon under study. Hence a man's height can be measured as being six feet, without reference necessarily being made to his 'tallness' – i.e. the fact that he is two inches taller than the norm. In contrast, a person's level of intelligence (IQ) cannot really be assessed without reference to the mean intelligence of the population.

Second, the strategy we adopt in measuring our variables should essentially be independent of how we define our concepts. There are certain costs associated with a lack of independence between them, one of which tends to be located in an over-dependence on operational definitions. As we saw in Chapter 1, defining our theoretical concepts by the way in which we measure them can be problematic. However, as with our IQ example, *many psychological concepts are defined in terms of variability*. This is particularly true in the area of 'individual' differences' (see Chapter 4).

Measurement strategy also influences the way in which we design our investigations and interpret our data. The use of variability in order to both measure and define psychological variables naturally requires testing, or experimenting on, large numbers of people in order to ascertain *norms*. One consequence of this approach is to shift the theoretical interest away from individual comparisons to group comparisons. In other words, in most of the designs and tests of significance discussed in later chapters we are interested in the 'average' behaviour of our samples. With the *t*-test, for example, we compare (population) means.

Johnston and Pennypacker (1980) refer to this particular measurement strategy (i.e. using variability) as the *vaganotic tradition*, and it is an implicit element in the conceptual framework of mainstream psychology. In a number of respects, the constraints of this tradition have prevented many psychologists from studying certain areas of interest, because such areas have been labelled subjective (e.g. the 'self'). Psychologists who have broken away from this tradition have often had to struggle to gain recognition. These sorts of problem may result partly from a traditional over-dependence on certain kinds of statistical tests for providing us with objectivity. In the following section we shall explore this further, as well as clarify some of the above points, by examining the constraints these statistical tests put on the researcher.

The Origin and Development of Statistical Tests and Their Consequences for Psychological Research

Although Galton collected a great deal of data, he did not have at his disposal the sophisticated data-analysis techniques we have today. It was Karl Pearson who began to develop statistical theory and tests in order to answer specific empirical questions arising from the study of psychology. Pure mathematics was a necessary but no longer a sufficient basis for the development of useful statistical techniques. In particular, Pearson was instrumental in the development of correlational techniques (e.g. Pearson's product-moment correlation coefficient – see Chapter 6). Subsequently, R. A. Fisher emerged as the major figure in the development of statistical techniques for use in experimental investigations. Fisher was one of the first to recognize the importance of the relationship between experimental and statistical design. This recognition grew partly out of a need for the experimental procedures to meet the underlying assumptions of certain statistical techniques, in particular those of analysis of variance. However, Fisher's techniques were specifically developed to deal with biological and agricultural data and their use can therefore place certain constraints on aspects of the psychological research process.

The sort of questions Fisher addressed himself to, and his source of data, are crucial to an understanding of the development of his techniques. In biology or agriculture the individual fruit fly or potato plant is of little concern. It is the making of inferences about populations from individual *random samples* which is important. For example, if you were testing the effect of a new fertilizer on the growth of a particular variety of corn, and successfully demonstrated its effectiveness, you would want to generalize its usefulness to the growing of all similar crops of corn. The crucial thing for the farmer is whether the crop as a whole is likely to benefit from the fertilizer, not whether a particular individual plant will. Thus, the type of statistical analysis developed to meet these needs does not say anything definite about individual characteristics – only about *group effects*. In adopting these sorts of technique, we are therefore restricted in the kind of inferences we can make from our data.

The form of statistical analysis we adopt may also influence the way in which we collect our data. For example, as we will see in the next chapter, one of the fundamental features of the classic experimental design is the notion of *random assignment* or *randomization*. This can be most readily achieved in a laboratory setting. The existence of well-established and convenient statistical tests, which have associated assumptions and restrictions, may thus constrain us to thinking only in laboratory terms. Such settings may not always be appropriate for the investigation of certain research problems. By limiting the scope and nature of the questions we can meaningfully ask, this in turn will restrict our theory-building. The development of alternative designs and types of statistical test have done much to release the researcher from the chains of his laboratory.

The Role of Significance Testing in Theory Construction

Neymann and Pearson were responsible for the development of the decision-making element of design and statistics. The essence of the decision-making process is to be able to quantify the probability that your findings were produced by 'chance', rather than being due to your manipulation of the variables under study. If you can establish that this probability is sufficiently low then the evidence you have collected can be interpreted as favouring your particular research hypothesis or question. A fairly arbitrary probability level ($P = 0{\cdot}05$) is traditionally chosen as a criterion, and this level is referred to as the *significance level*. If the probability associated with your particular finding falls below this level it is considered to be significant and if above, it is considered non-significant. The statistical relevance of testing the null hypothesis, which is the correct term for the above description, will be discussed more fully in Chapter 5. For now, we will explore the notion of whether support for your hypothesis should be seen as unequivocal support for your theory, i.e., whether tests of significance are tests of theories. The arguments are complex but you do need to have an understanding of them in order to be able to design your own investigations and build theories with confidence.

First, there is the danger that a significant result can be perceived as an end in itself. In other words, a significance test is regarded simply as a test of a theory, and not as part of the process of theory construction. Equally, a non-significant result is often perceived as the demise of a theory. In Fisher's terms, a non-significant result should simply warn the researcher to 'reserve judgement'. After all, as we saw in Chapter 1, failure to reach significance may not necessarily be a reflection on the adequacy of your theory. We also discussed the importance of the distinction between theoretical and statistical significance. If your theory is unsubstantial or even trivial, then finding a statistically significant result is not going to improve your theory. Equally, consider a situation in which you and 99 other researchers had a really interesting theory, and carried out identical well-designed experiments to test hypotheses generated from that theory. If all of you chose a significance level of $P = 0{\cdot}05$ (i.e. a 1 in 20 chance), then five of you, just by chance, would obtain significant results. As suggested in Chapter 2, one or two of these five are likely to get their findings published!

A further consequence of using the kinds of investigative design outlined in this book is that, *as you increase your sample size, you increase the likelihood of finding a significant result* (why this happens will be discussed later). So, if you want to ensure that you get some significant results in your dissertation or project, then a simple but time-consuming tactic would be to run hundreds of subjects! Large sample sizes provide the best conditions for the testing of precise hypotheses (i.e. in an explanatory strategy), but they can create problems for explorative investigations. In a correlational study, for a sample of 700 persons you only need a correlation coefficient of 0·08 to be significant at the $P < 0{\cdot}05$ level. Again this highlights the danger of thinking only in terms of statistical significance, at the expense of theoretical importance. Whatever your sample size your chances of obtaining a significant correlation are the same, but in an explorative strategy you are looking for more substantial associations (i.e. large correlation coefficients), in order to indicate those variables worthy of future investigation. It may be advisable to select a stricter significance criterion (perhaps $P < 0{\cdot}001$) or to run fewer subjects (perhaps only 50). If you don't, you may find yourself with many significant findings which are both unexpected and difficult to interpret.

Finally, the single-minded attitude of many psychologists to tests of significance has led to an over-emphasis on the use of inferential statistics to the cost of descriptive statistics. This has resulted in a bias towards the use of explanatory research strategies. However, many researchers have argued that valid explanation needs to be built on a foundation of exhaustive description: in other words, *before you attempt to explain something you must describe it.* Without such a foundation you may make errors in problem selection and in the making of inferences from your results.

Constraints Relating to Choice of Statistical Test

Given that our dominant measurement strategy revolves around the concept of variability, it is hardly surprising that many of the statistical tests used in psychology are themselves based on analysing variability (variance). However, as we have seen, many of the techniques that we use were developed by researchers from other disciplines, and hence for a different purpose. In an ideal world we should develop an appropriate statistical test to answer our specific research question. However, if we had to design new statistical tests for almost every set of data collected, we would have very little time or energy to collect the data in the first instance! Also very few of us have the statistical expertise to devise our own tests, or easy access to those who can. Consequently, there exists a pool of statistical techniques which psychologists draw upon when analysing their data.

As previously pointed out, the practice of using statistical tests which have not been specifically designed for our particular purpose can have an obvious effect on our general theory-testing and construction. It can also result in more specific constraints on the actual conduct of the research and on the kind of variables we can study.

The range of available statistical tests can be divided into two major categories: *parametric* and *non-parametric*. The conduct of the research process is often governed by the need to fulfil the assumptions associated with parametric tests, because of the benefits connected with their use. Parametric tests are those which are limited by a necessary adherence to certain underlying assumptions, mainly concerned with the characteristics of the population distribution from which the sample was drawn (see Chapter 5). Non-parametric tests make less stringent demands regarding the characteristics of your population and are therefore often referred to as 'distribution-free' tests.

The category of test you wish to use can limit your choice of measurement scale (nominal, ordinal, interval and ratio), and vice versa. Parametric tests normally require data measured on at least an interval scale. Non-parametric tests usually require variables to have been measured on a nominal or ordinal scale (although it is relatively simple to convert interval or ratio scale data into ordinal scale data by ranking the scores). The questions we raised in the early part of the chapter, regarding the appropriateness of interval scales for measuring psychological variables, become critical in the choice of statistical test. There are at least two main assumptions that psychologists make to justify their use of interval scales. First, we often assume that the variable being measured is normally distributed in the individuals being observed. We can thus have a situation of the researcher manipulating the units of his scale until they match the assumed normal distribution of the variable under observation. Our earlier example of Galton's work on the measurement of intelligence is a prime instance of this 'intuitive' creation of an interval scale. Second, a constant unit of measurement is essential to an interval scale. Many psychologists therefore make the assumption that giving a correct, or affirmative, response to any one item of a test or questionnaire is exactly equivalent to giving the same response to any of the other items or questions. Certain procedures have been developed to allow an investigator to make these assumptions, and these will be discussed in Chapters 5 and 8.

The above points emphasize how important it is to *ensure that your measurement scale and chosen form of statistical analysis are compatible with each other, and with the aims and objectives of your particular investigation.*

A preference for the use of either a parametric or a non-parametric statistical test in a particular investigation can also influence your research design and procedure. The restrictions on the research process connected with the choice of a parametric test would not be so prohibitive if it wasn't for their considerable advantages, in terms of design scope (size and complexity) and flexibility of use (see Chapters 5 and 9). Also, we have already discussed the way in which the experimental logic underlying the statistical procedures developed by Fisher requires the random sampling and assignment of subjects. Not surprisingly, the statistical tests associated with Fisher are examples of parametric tests, and random sampling and assignment is one of the assumptions associated with the use of this category of tests. Thus the category of test you choose can determine the way in which you select your subjects. However, there are occasions when subjects cannot be randomly sampled or cannot be randomly assigned to conditions. The breaking of this assumption, therefore, *may* restrict your choice of statistical test to those contained in the non-parametric category. However, in investigations which involve 'individual-difference' (organismic) variables such as sex, intelligence or personality, although subjects can be randomly sampled they cannot be randomly assigned. In some of these situations the use of parametric statistics has become common practice, but this does tend to introduce problems regarding the interpretation of results (see Chapter 4).

Finally, your choice of test (parametric or non-parametric) can constrain the nature of your results and their interpretation.

Parametric tests are more 'powerful' than their non-parametric equivalents when applied to the same data. This means that parametric tests have a greater probability of correctly rejecting the hypothesis that your results have been produced by chance. The 'power' of these tests is one reason for people's preference for them. However, remember that their reliability is dependent on their assumptions not being breached. If they are breached, this may increase the likelihood of your making an incorrect decision in rejecting the possibility that your results are simply a reflection of chance variation (more fully explained in Chapter 5). Another characteristic of parametric tests is that they are more specific. They can answer direct questions about the comparisons of mean scores and thus aid your interpretation. Non-parametric tests can compare groups (populations), and therefore compare means (population means) indirectly, but along with their added design scope and flexibility, parametric tests frequently provide you with a more useful, precise and sharp tool for dissecting and interpreting your data.

Further detailed information on the differential nature of these two categories of tests, and the factors which may determine your choice of one or other of them, will be found in Chapter 5.

Summary

This chapter explored the general issue of measurement in psychology and described the nature and use of four types of measurement scale. It also discussed psychology's attempt to deal with some of the problems of measurement through the adoption of a particular measurement strategy. The remainder of the chapter focused on the origin and development of certain kinds of statistical test, and the difficulties and constraints associated with the application of some of these tests in psychological research. Choice of statistical test, and in particular the choice between a parametric and a non-parametric test, was shown to have an important influence on the way we measure our variables, and on how we collect, analyse and interpret our data. A more detailed discussion of the advantages and disadvantages of the two major categories of statistical test will be presented in Chapter 5.

The most important message contained in this chapter was to *plan ahead*. Without a clear plan, you may end up with data in a form which cannot be appropriately or effectively analysed. Your research question and your choice of research format, data-collection technique and measurement scale may all affect the range of statistical tests available to you. Equally, your form of data analysis is likely to influence the sort of questions you can answer, and hence the interpretation of your data.

4 Correlational and Experimental Research Formats

This chapter describes the two main research formats used to collect data in psychology: the experiment and the correlational study. As we shall see, each of the two research formats has advantages and limitations in terms of the types of inference that can be drawn from the data obtained. We shall also see that researchers tend to use one or other of these formats to the exclusion of the other. Indeed, Lee Cronbach (1957, 1975) has labelled *experimental psychology* and *correlational psychology* the 'two disciplines of scientific psychology'.

The Simple Experiment: Active Observation

The experiment is concerned essentially with the location of causation. The experimenter has control over an *independent* variable and looks for changes in a *dependent* variable. Notice that the presumed cause, the independent variable, temporally precedes any effect on the dependent variable. For instance, you may wish to see if a noisy environment (the independent variable) has a detrimental effect on performance of a skilled task (the dependent variable). You could put some of the subjects in your experiment in a noisy room and some in a quiet room and measure differences between the two groups' performance on a chosen task. Obviously, noise can be produced in different ways – with a pneumatic drill, for example, or by people arguing – and at different intensities. Similarly, the skilled task could be a cognitive skill, such as multiplying numbers together in your head, or a physical skill such as balancing coins on their edges. Indeed, a single dependent variable can also usually be measured in several ways. For instance, the cognitive skill of multiplying numbers together could be measured as the number of errors made *or* the number of correct answers given within a certain time period (these two measures of the dependent variable would not necessarily be strongly associated) *or* the number of correct answers divided by the total number of attempts at answering. Clearly, your choice of independent and dependent variables and the way in which you measure the dependent variable should be closely linked to the theory underlying your experiment and the type of generalization you may wish to make. For instance, it is arguable whether any of the results of any of the experimental designs outlined above would help us in deciding whether listening to a stereo cassette recorder played loudly in a car affects driving performance!

In our imaginary experiment, in order to see if presence or absence of noise affects performance we must test the performance of subjects on a skilled task under a noise condition against the performance of subjects on the same task under a no-noise condition. The no-noise condition is called the *control condition*. Subjects performing under the control condition should be performing under *exactly* the same circumstances as those under the experimental condition, *with the exception* that they are not subject to noise. In this way, if a significant difference in performance is observed between the two conditions we infer that it could only have been *caused* by one thing, the presence or absence of noise.

By 'exactly the same circumstances' we mean two things: (1) that the experimental environment and task circumstances are equivalent, and (2) that the subjects performing under each of the conditions are also equivalent or exchangeable. Obviously, (1) is easier to establish than (2). Experimenters attempt to achieve subject equivalence or exchangeability by one of two main methods, either by *matching* or by *randomization.* Matching involves pairing each of the subjects in the control group with a subject in the experimental condition. The two subjects are matched on all, or at least all the known, factors which could influence the relationship between the independent and dependent variables. For instance, if partially deaf people were allocated predominantly to the noise condition in the previous example, observed equality of performance between the noise and the no-noise condition would perhaps be anticipated!

To illustrate the point further: suppose an experimenter was interested to see if a new method of teaching affected subsequent examination grades. If the new method of teaching was to be tried on one group of students, a control group would obviously also be needed. It would consist of students who would continue to be taught by the old method. In this instance it would seem sensible that the two groups of students should be matched in terms of age, IQ, previous academic performance and so forth, since differences in these variables could, by themselves, affect subsequent examination grades. Unfortunately, other variables such as achievement motivation may *not* have been thought of by the experimenter and *not* used in the matching process, and yet may contribute to differences between the groups on the measure used as the dependent variable. For this reason matching can never control for *all* non-equivalence between groups of subjects. Once the subjects who are to take part in the experiment have been matched, each pair should then if possible be allocated to the control and experimental conditions *at random*: in other words, each of the two subjects should have an equal chance of being selected for a particular group. Random allocation prevents unmatched factors from having any systematic effect on the outcome of the experiment. Matching and randomization are thus two important elements of an experimental design, which help rule out possible alternative explanations of the observed effects on a dependent variable.

In summary, the *ideal* experiment is a tightly controlled study in which the effects on the dependent variable of changes in the independent variable can be assessed and the range of possible explanations as to the cause of any observed effects can be reduced to a minimum. It can then be inferred that any difference between the control and the experimental group on the dependent variable is the result of the effect of the independent variable and/or chance variation. As we shall see in Chapter 5, statistical tests allow us to infer the extent to which a difference is due to the manipulation of the independent variable, rather than to chance.

Other Experimental Designs

Up to now we have been talking about situations in which there is only one experimental group, together with one control group from whom the independent variable is withheld. Of course there may be several experimental conditions, or *treatments*, one for each of the levels of the independent variable – for instance, low, medium and high noise. In a *factorial experimental design* two or more independent variables are manipulated *at the same time*. Each level of the first independent variable is combined with each level of the second independent variable. The extent to which each is responsible for changes in the dependent variable can thus be evaluated. Indeed, the possibility of an *interactive* effect of the two independent variables can be studied. For example, a decrement in skilled performance caused by noise and flashing lights may be much greater when these 'distractors' occur together than when they occur singly. The advantages of factorial and related designs over the simple experimental design are discussed and illustrated in Chapter 9.

As already pointed out in Chapter 3, it is sometimes inappropriate to assign subjects randomly to control and experimental conditions and instead what have been called *pre-experimental* designs are often used. A commonly used design of this sort is the *repeated measures design*. Here one group of subjects serves both as control and experimental group. Observations are taken on the dependent variable before the treatment is implemented and then again after the treatment. For instance, the number of students absent from a certain lecture course could be measured before and after a change in lecturer halfway through the course. However, a causal inference that the introduction of the second lecturer produced more student attendance is open to counter-interpretations – the second half of the course is nearer to the exams and perhaps contains revision material! Fortunately, in other instances the order of presentation of treatments to individual subjects can often be successfully counterbalanced or randomized. Here the repeated measures design is experimental. See Colin Robson's *Experiment, Design and Statistics in Psychology*, Chapter 8, and also Spector (1981). Chapter 9 of our book develops the discussion of experimental design within the context of the analysis of variance.

The Simple Correlational Study: Passive Observation

The simple correlational study is concerned essentially with identifying, or measuring the strength of, association or correlation between two variables. In its simple form the correlational study can say little about what caused what. For instance, you might find that aggressive behaviour is positively related to lack of close friends. If so, does aggressive behaviour cause lack of friendship, does lack of friendship cause aggressive behaviour or is there a third factor underlying both these observations? Often one of the variables to be correlated precedes the other temporally. In this case the causal variable may seem obvious. For instance, if lack of motherly love was correlated with vandalism in later life it is tempting to infer that the former 'caused' the latter. However, vandalism in later life may be more associated with unemployment than with maternal deprivation. Clearly, then, inferences about causality from obtained associations or correlations should be made with caution.

In correlational studies, the researcher *cannot* manipulate an independent variable and observe the resulting effect on a dependent variable. The researcher simply *observes passively* the relationship between a pair, or pairs, of variables. Thus correlational studies do not have control groups, nor can the researcher randomly assign subjects. Indeed, dependent and independent variables cannot be distinguished. For these reasons a single correlation coefficient (more about this later in Chapter 11) cannot be seriously entertained as an index of causality. In fact, as in our aggression example, three possible but contradictory hypotheses can be put forward to explain an obtained correlation between two variables, say X and Y: (1) X may cause Y; (2) Y may cause X; (3) the obtained relationship between X and Y may be caused by a third variable.

Causal inferences are, however, often drawn from correlational data. The validity of such an inference depends on the satisfaction of three commonly accepted conditions. First, there must be evidence of *time precedence*: for X to cause Y, X must precede Y in time. Second, there must be evidence of a *relationship*: the two variables must be, to some extent, associated or correlated. Third, the association or correlation must be *non-spurious*: there must not be a third variable that causes both X and Y such that once its effects are controlled the obtained relationship between X and Y disappears. For instance, height may be correlated with scores on a test of statistical ability, but this correlation may be due to the relationship between height and age! Methods of partial correlation can 'partial out' the effects of age to see if a correlation still exists between two variables (height and statistical ability) independent of a third variable (a person's age). These methods are elaborated in Chapter 11.

In summary, the simple correlational study is the *passive observation* of relationships between variables. In most cases Nature has already implemented the independent variable and the researcher can do little more than make an educated guess as to what the independent variable was, or is.

Comparison of the Experiment and the Correlational Study

While the experimenter is interested in self-created variation, the correlator is interested in the variation already existing between individuals, and/or groups of individuals, within or across cultures. Thus the correlator's aim is to reach an understanding of Nature's own experiments.

Conversely, to the experimenter individual differences are often seen just as a problem, rather than as intrinsically interesting. A large amount of variation obtained between subjects within a treatment is seen as an indication that the treatments are not controlling behaviour tightly enough. Individual variation is termed *error variance* and it is the one source of variance the experimenter tries to reduce, by matching subjects or using homogeneous subject groups.

Since the correlator's main interest is individual variation, heterogeneous samples are often used in order to produce the maximum possible variation between people. As we shall see in Chapter 7, the correlational approach leads directly to the study of psychometrics, or the measurement of individual differences.

In research undertaken within a correlational research format, no attempt is made to change the extent to which an individual has a certain behaviour, disposition or capacity. Variation is produced by selecting individuals who differ, or who are expected to differ, on the property of interest. The experimenter, on the other hand, obtains variation by subjecting a given group of subjects to different treatments. It follows from this that *experimental variables* and *correlational variables* can be distinguished. For example, an experimenter may attempt to manipulate the independent variable 'frustration' by giving subjects various tasks of differing degrees of difficulty whilst telling them that all the tasks are relatively easy. Conversely, the correlator may use a questionnaire designed to measure the variable 'frustration tolerance', which may contain questions such as 'do you often feel frustrated and angry when you cannot complete a simple task correctly?' The correlator may subsequently find that there are differences between individuals on this variable.

It should now be clear, from our discussion here and in Chapter 1, that the experiment is the primary research format used within an *explanatory research strategy*. The correlational format tends to typify the *descriptive research strategy* whilst the *exploratory re-*

search strategy often combines the investigative approaches of the experiment and the correlational study. *Data collection* in the experimental format tends to concern measures of performance (e.g. number of words recalled) under changing task conditions, while the correlational format tends to involve data collected by surveys, interviews or questionnaires. *Data analysis* in the experimental format looks for differences in performance under different conditions whilst the correlational format uses the statistics of correlation and association. The *theoretical perspective* of the experimental format can be characterized as a search for 'general laws' of behaviour that apply to all people in similar situations, and that of the correlational format as the measurement and interpretation of differences between individuals or groups of individuals.

Throughout this chapter we have contrasted the correlational or passive observational research format with the experimental research format. However, *correlations in the statistical sense can be used to analyse data from experiments as well as data from passive observational studies*. For instance, in our imaginary experiment to investigate the effect of a noisy environment on the performance of a skilled task (p. 49), a questionnaire measure of 'frustration tolerance' could be given to those subjects performing within the 'noise' condition. Individual differences in skilled performance, if they have not been totally eliminated by the experimenter's matching, could then be correlated with individual differences in frustration tolerance. Perhaps people who are tolerant of frustration perform best on a skilled task under annoyingly noisy conditions! Similarly, *statistical techniques involving measurement of differences can be used to analyse data from passive observational studies.* For instance, scores on a questionnaire measure of individual authoritarianism could be compared between two samples, taken respectively from the populations of an 'authoritarian' and a 'democratic' state.

However, we believe that measures of individual differences should not be used as independent variables in experimental designs. For example, you may think of categorizing a sample of people as either high-scale-scoring authoritarian or low-scale-scoring authoritarian, and using this dichotomy as an independent variable. But a measure of individual differences is not an independent variable, and does not become one by categorizing people's scores and treating the categories as if they represented a variable under experimental control. This is because it is impossible to assign people at random to levels of the 'treatment' and because levels formed in this way will differ from each other on *all* individual difference variables correlated with the one categorized. In other words, there may be an illusion of control of variables and a tendency to place inappropriate causal interpretations on significant findings. Also, if several individual difference measures are categorized to form additional 'independent' variables in a factorial design, their effects on the dependent variable will be distorted if the individual difference measures are to any extent intercorrelated.

An important additional point is that partitioning an interval-level variable into categories is a transformation that involves a *loss* of information. We feel that it would be much more appropriate to correlate the scores on the individual difference measure with the scores on the 'dependent' variable.

Real-world Validity of a Research Design

Real-world or ecological validity concerns the extent to which the results of an experiment or correlational study can be generalized to other situations and other people, i.e. to the world outside of the particular people and the particular task studied. Two sorts of questions concerning real-world validity that you should ask yourself are:

1. If the subjects used in your study are students sampled from a particular college, do you expect your results to generalize to other student groups, or indeed to people in general?

2. Are the levels of the independent variable you manipulated applicable to the real world?

For example, you may have investigated how college students make choices between risky and less-risky gambles with imaginary payoffs. But what if the money were real, and the setting a casino rather than a psychological laboratory?

As we discussed earlier, in order to design a well-controlled experiment where attributions of causality to changes in an independent variable are relatively indisputable, the experimenter will tend to use artificial laboratory tasks and homogeneous subject groups. In order to produce results that have meaning for the real world, however, the experimenter often either chooses or is forced to perform experiments in complex, less tightly controlled 'realistic' settings with more heterogeneous subject groups. Consequently, competing attributions of causality are possible and, indeed, likely. Such realistic settings are, as we have seen, a major interest of the correlator! *Field experiments* in, for example, schools or places of work are a compromise between laboratory experiments and passive observation. To some extent, they circumvent the problem of the artificiality of the laboratory.

Let us end this chapter with a quotation from Dalbir Bindra and Ivan Scheier (1954) who note that researchers working within the correlational and experimental research formats *may* have something in common!

> If the psychometric researcher and the experimentalist agree on anything, and there is some doubt about this, it is that the other kind of psychologist plays in another league (class B).

We will have much more to say about a reconciliation of experimental and correlational research formats in Chapter 12. In the intervening chapters we will detail the methods and techniques of Lee Cronbach's 'two disciplines of scientific psychology'.

Summary

This chapter compared and contrasted the simple experiment with the simple correlational study involving passive observation. The researcher working within the experimental format was shown to be interested in self-created variation, while the researcher working within the correlational format was seen to seek an understanding of Nature's experiments. The ideal experiment was shown to create the best conditions for inferring causation. However, well-controlled experimentation can produce results that are difficult to generalize to situations and people outside the psychological laboratory. Conversely, field experiments, conducted in real-world settings, tend to result in competing attributions of causality, a major characteristic of most correlational studies.

5 Some Basic Concepts and Assumptions Underlying Statistical Tests

In Chapter 3 we stressed the importance of the concept of variability in terms of the measurement strategies and kinds of statistical test used by psychologists. We shall now discuss in more detail the role of variability in statistical analysis, and explore some of the concepts and assumptions underlying statistical theory, statistical decision-making and statistical tests. We hope that the information will not only be of some intrinsic interest, but will also provide some insight into the statistical tests dealt with in later chapters.

Important Statistical Concepts

1 Inferential and descriptive statistics

First, we need to review the distinction between descriptive and inferential statistics. As indicated in Chapter 1, descriptive statistics help us describe data, whereas inferential statistics help us explain a phenomenon by drawing inferences from our data. In statistical terms, *descriptive statistics describe samples and populations; inferential statistics help us make inferences from samples about populations.* The statistical tests discussed in later chapters are all of an inferential nature, and offer a means of assessing the probability that the inferences drawn from our samples are incorrect. An understanding of inferential statistics requires some knowledge of samples and populations, and in order to understand the basic mechanics underpinning these kinds of statistical method, you also need to know something about sampling and the existence of various kinds of distributions.

2 Samples and populations

Inferential statistics permit us to make inferences from samples about populations, i.e. to make generalizations. Let us therefore define what we mean by samples and populations.

A population must include every case or member which you have defined as being of interest: in other words, it contains the total group you wish to generalize to. For example, a population may contain all the children attending comprehensive schools in Britain. Populations tend to be, but do not have to be, rather large. On the other hand, *a sample is a subset or a relatively small group selected from a population*, and we generally hope that our sample will be representative of the target population. If it is representative, then the behaviour of our sample will reflect the behaviour of the population as a whole. Thus, if we are to make realistic generalizations then we must be sure that our sample is truly representative. Obviously, making inferences on the basis of a biased sample should be avoided.

The various sampling procedures are methods of selecting samples which attempt to ensure that they are not unduly biased. One method of unbiased sampling is to make sure that you know the probability of occurrence of any particular sample. Samples selected on the basis of this technique are called *probability samples* and they have characteristics which have important implications for the making of inferences from the sample. There are different types of probability samples, but the one that we shall focus on here is the *simple random sample*, which is the type most frequently used in psychology.

True random sampling makes two demands on the researcher. First, *each individual must have an equal chance of being selected*, and second, *the selection of any one individual must not bias the selection of any other*. In sum, these demands should ensure that individuals are selected *independently* of each other. To satisfy the second demand fully you should sample 'with replacement'. If you do not replace the selected individual back into the population, then the population size has been reduced. Reducing the population size effectively increases the selection chances of any one of the remaining individuals. However, with large populations and relatively small samples, sampling without replacement has an insignificant effect on the independence of future selections. In general, *as long as the sample size is no greater than one-fifth of the total population, then the effects which may arise from sampling without replacement can safely be ignored.* Therefore most simple random sampling involves sampling without replacement.

3 Frequency distributions

If we were presented with the problem of determining the mean intelligence of all children attending British comprehensive schools, then we could measure the IQ of every single child from that population. The number of children achieving each IQ score could then be plotted on a graph to give us a distribution of the IQ frequencies. Figure 5.1 is such a frequency distribution, with a continuous curve superimposed on the hypothetical distribution obtained. In our example, the most frequent score (the *mode*) is also the same as the mean and the median. Notice how scores above and below the mean become more infrequent, or rarer, as the IQ scores increase or decrease in size. This factor gives the distribution a very characteristic bell shape. In this particular example, we have produced what is called a *population distribution*. The two important measures used to describe population distributions are their mean and standard deviation, and these are referred to as the *parameters* of

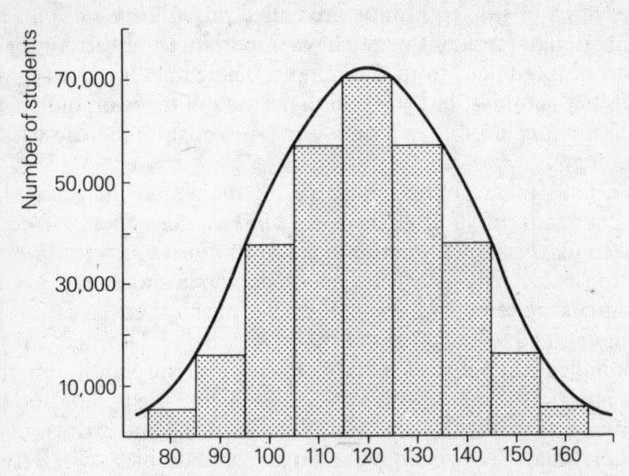

Figure 5.1 A hypothetical population distribution of IQ scores of British comprehensive school children

our population distribution. Of course, if we had actually undertaken the exercise outlined above in order to determine our population parameters, then we would not require a sample and there would be no need to make inferences: in other words, our parameters would provide us with all the information we would need.

However, to test every individual in a population would clearly be a very arduous and time-consuming task. Instead, we could draw a random sample and make inferences from it about the general population from which it was drawn. The sample will of course have its own frequency distribution (see Figure 5.2). Where samples are concerned, statistics like its mean and standard deviation are called *sample statistics*, and the frequency distribution they describe is called a *sample distribution*. The mean and standard deviation of a sample are represented by the roman letters '$\bar{X}$' and 'S' respectively. From Figure 5.2, you will notice that although the distribution is of a similar bell shape to Figure 5.1, its mean and standard deviation are different from the parameters of its population distribution. Indeed, the only way to ensure that the mean and standard deviation of a sample are equal to those of its population is to sample the whole population! The question then arises: if our sample statistics are different from their population parameters, then *how representative of that population is our particular sample?*

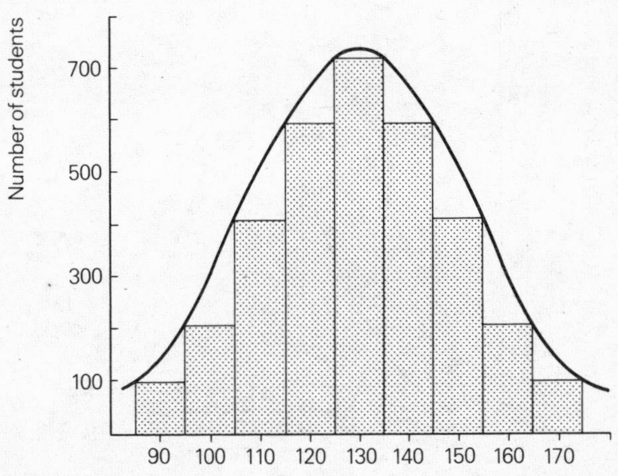

Figure 5.2 A hypothetical sample distribution of IQ scores of British comprehensive school children

If, instead of drawing one sample, we drew a vast number of samples then we would be in a better position to answer this question. We could then plot the frequency distribution of the means for all our samples and that would give us information about the variability resulting from taking a large number of samples of the same size. Notice that the distribution (Figure 5.3) is again bell-shaped, and that *the mean of the means is equal to our population mean*. If we were able to draw up such a distribution then it would provide us with interesting information about the likelihood, or the frequency of occurrence, of a sample producing any one particular mean. In the case of our school children's IQ distribution, to have drawn a sample with a mean IQ of 100 would have been a very likely event, so that the sample could be considered representative of the population. On the other hand, to have drawn a sample with a mean of 135 would have been an improbable event, and therefore the sample could not be considered representative of the population.

Obviously, to take a very large number of samples is again impractical. Instead, we could develop a *theoretical distribution* drawn up through the application of mathematical theory. This would represent the distribution that would have been obtained if

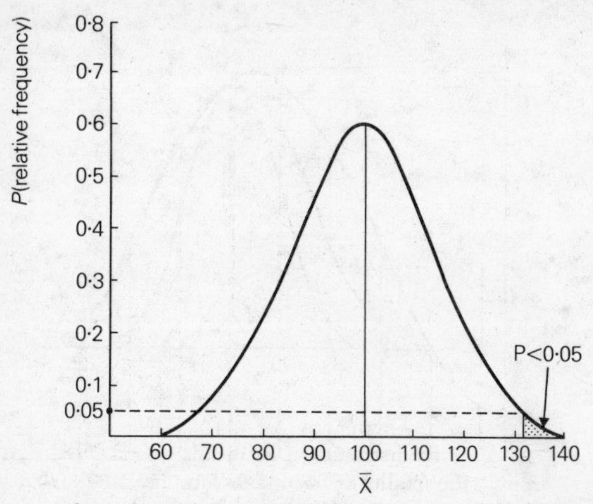

Figure 5.3 A hypothetical sampling distribution of means (IQ scores of British comprehensive school children)

we had taken an infinite number of samples. Such a distribution is called a *sampling distribution*. For our particular example, if we wished to assess the representativeness of our sample mean, we would need to refer to a *sampling distribution of the means*.

There are several important characteristics of sampling distributions. First, as we have already noted, *the mean of the sampling distribution is equal to the mean of the population* (the standard deviation is not, in general, the same as the population standard deviation). It is therefore desirable that the mean of any one sample should be as close as possible to the mean of the sampling distribution, and hence the population. Indeed, the less variation there is between sample means the more representative any one sample mean will be. To reflect the reduction in sampling error that results from reduced variance, *the standard deviation of a sampling distribution is called the standard error*. A small standard error would mean that there is a higher probability that any one of your sample means would be a good estimate of your population mean (see Figure 5.4). This leads us on to the second important characteristic, which

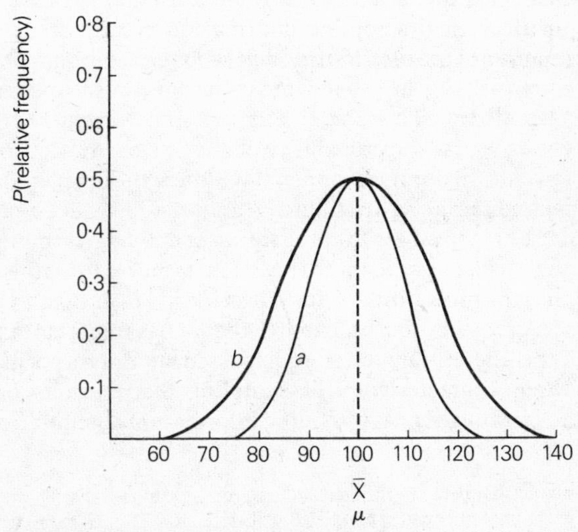

Figure 5.4 Two sampling distributions of means (*a* and *b*) with different standard errors (SE); the smaller SE of *a* makes it more probable that any one sample mean is a better estimate of the population mean than a mean drawn from distribution *b*

is that *the standard error of a sampling distribution decreases as the sample size increases.* The size of your sample should not be confused with the number of samples taken. If the sample size is increased or decreased the shape of the associated sampling distribution changes. Therefore, a number of distributions can exist – depending on sample size, each sampling distribution consequently referring to an infinite number of theoretical samples of the same size. To recapitulate, the larger your sample size (or N size), the smaller the standard error associated with its distribution is, and therefore the closer your sample mean will be to the population mean.*

A third point to note about theoretical sampling distributions is that they are represented by a continuous curve without gaps or breaks – see Figure 5.3. The associated assumption is that any variable that is represented by a theoretical sampling distribution should be continuous rather than discrete. This means that it should be possible to achieve any value of the variable (e.g. reaction times). Social class would be an example of a variable which is discrete rather than continuous. The continuous nature of the curve representing a theoretical sampling distribution means that *the area under the curve of the distribution can be expressed as a probability (relative frequency).* In other words, the total area under such a curve is equal to 1. This characteristic is most useful, because it allows you to assess the probability of any one event occurring in a group of events. In our IQ example, *if* we knew the characteristics of the associated sampling distribution, and we drew just one sample, we would be able to assess the probability of drawing a sample with a mean equal to, or larger than, that particular mean. In Figure 5.3, we can see that the probability of drawing a mean larger than 135 is less than 0·05. As we will see, probability theory has important implications for the inferential processes discussed later in this book.

The three characteristics of sampling distributions outlined above can be thought of as providing information about the *location*

* The relationship between the standard error of the sampling distribution and the standard deviation can be stated mathematically as:

$$SE = \frac{\sigma}{\sqrt{N}}$$

and thus demonstrates why SE decreases with an increase in N (see also the discussion of the Central Limit Theorem on p. 68).

(the mean), the *variability* (standard error) and the *shape* (the curve) of a sampling distribution.

4 The normal distribution

The normal distribution is one example of a theoretical distribution, and its mean and standard deviation (parameters) are symbolized by the lower-case Greek letters mu (μ) and sigma (σ) respectively. Notice how similar the normal distribution (Figure 5.5) is to our other particular examples of sampling distributions: they are all bell-shaped. However, the shape of the normal distribution has other important characteristics. First, *it is symmetrical about the mean*, one consequence of this symmetry being that the mean, mode and median are all represented by the same point on the distribution. Second, the ends of the curve never quite touch the horizontal axis, that is, *the distribution is asymptotic*. This means that there is a very small chance of getting an extremely low or an extremely high score. Third, the *normal curve is continuous*. This, as we already noted, implies that variables measured by reference to a normal distribution must be of a continuous rather than a discrete nature. These characteristics will apply to any curve which has been generated from the normal distribution equation. Although this is

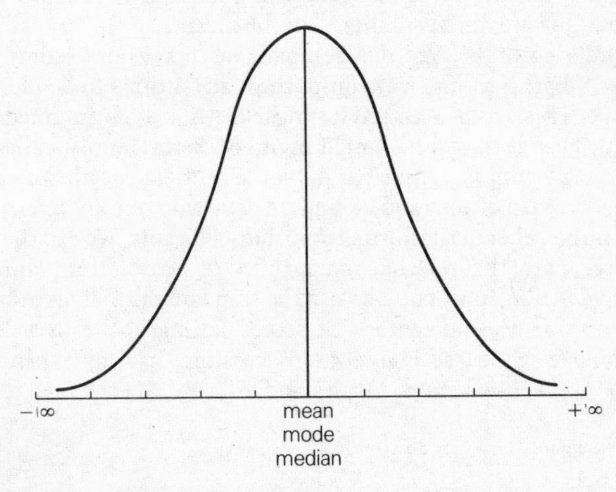

Figure 5.5 A normal curve or distribution

not the place to discuss the actual equation, it is useful for you to realize that you need to know only the mean and standard deviation of any group of scores (plus two constants) in order to draw a normal curve. In fact, *the normal distribution represents a family of distributions* whose members differ in shape according to their means and standard deviations. However, although the shape of the curve may change, its essential characteristics do not.

Psychologists are often concerned with the position of a particular score (or individual) relative to other scores in the distribution, as in IQ assessment. So, for example, if we knew the mean and standard deviation of a set of scores, we could express the position of one score in terms of how many standard deviations it is away from the mean. When a score is expressed in these terms it is referred to as a *standard score* or *z-score.** All distributions of *z*-scores have a mean of zero and a standard deviation of 1. Scores like these are particularly useful when you wish to compare an individual across a number of distributions, each with a different mean and standard deviation. For instance, if an individual undertook three aptitude tests (A, B, and C) and achieved standard scores of +1, +2 and −1 respectively, then we are able to say that he did better than average on Test A, very much better than average on Test B and worse than average on Test C. Thus, the magnitude of the *z*-score tells you how many standard deviations a score is from the mean, and the sign (+ or −) tells you whether it is greater or smaller than the mean.

In order to make *rigorous* comparisons between *z*-scores from different distributions, we require the distributions to be identical in shape. This can be ensured for those distributions assumed to be normal, by referring to a single member of the family called the *standard normal distribution.* As you might expect, this distribution has a mean of zero and a standard deviation of 1. By referring to only one member of the normal distribution family, we can then not only compare the relative magnitudes of scores from different normal distributions, but also use the standard normal distribution to determine the percentage of scores falling above or below a particular *z*-score (see Figure 5.6). Therefore, having transformed your original observations into a standard score you can then refer to

* The formula for calculating a *z*-score is:

$$z = \frac{X - \bar{X}}{S}$$

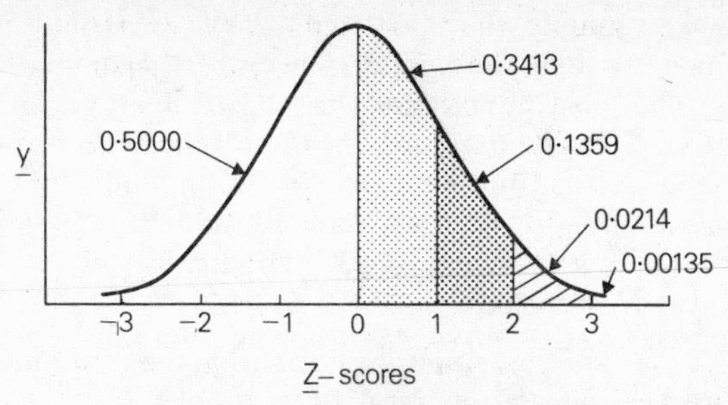

Figure 5.6 Some areas underneath a normal curve for different *z* values

the normal distribution tables (see Table I in the Appendix) to ascertain the proportions of area under the standard normal curve that correspond to different values of *z*.

It is important to note that a *z*-transformation can be undertaken on any score from *any* distribution. However, it does not change the shape of the distribution, so only those scores from distributions which approximate to the normal should be compared by means of the standard normal distribution. *A z-transformation will not normalize your distribution.*

In addition to its use for standardizing scores, the normal distribution has other practical applications. It can be used to represent the distribution of many psychological variables. This is because uncontrolled-for variation (or error) in any system actually approximates to a normal distribution, a fact referred to as the *normal law of error*. For this reason one finds many variables in nature that approximate to a normal distribution: height and intelligence are two examples. In these cases, variables not directly studied by the researcher, such as genetic endowment or the quality of the environment (e.g. diet), may well contribute to a person's height and IQ score. Although such extraneous variables will contribute to the individual variation, this variation will itself be normally distributed. This fact becomes extremely important for psychologists, because most of the variables that we are interested in are made up of a combination of variables.

Finally, the normal distribution is important theoretically because, via mathematical theory, it can help answer the question of how we produce a theoretical sampling distribution. Given that we have drawn a sample, the associated sampling distribution may be determined through the application of particular mathematical theories. The most important of these is called the *central limit theorem*. As already mentioned, the three important characteristics of sampling distributions are its mean, its variability and its shape. The central limit theorem can provide us with the information necessary to draw up a sampling distribution by stating that *as the sample size (N) increases*:

(*a*) The mean of the sampling distribution tends to equality with the mean of the population (mean or location).

(*b*) The variance of the sampling distribution becomes equal to the population variance divided by the sample size (σ^2/N) with a standard deviation of $\sqrt{(\sigma^2/N)}$ (variability).

(*c*) The sampling distribution approximates to a normal distribution (shape).

The theorem is thus important because it can provide us with the mean, the variance and the shape of any given sampling distribution. Therefore, the useful characteristics associated with the normal distribution can, in special circumstances, be applied to these sampling distributions. The necessary conditions for the legitimate application of these useful characteristics will be discussed later in the chapter.

5 Estimation and hypothesis-testing

Up until now, we have dealt with samples in terms of ascertaining how far a sample is representative of its general population. In the case of our IQ example, the sample statistic of interest was the mean, and we were concerned with whether our sample mean was a good and an unbiased estimate of the population mean. Such estimations are not restricted to the mean. We can also make other estimates from our samples, such as the standard deviation of a population. Thus, by taking a sample we hope to learn something about the parameters of our population, i.e., our sample statistics are used to estimate our population parameters.

Sample statistics can provide us with the information necessary to make a *point estimation* of our population parameters. However, because of sampling variation (or error), it is more sensible to talk in terms of an *interval estimation* as an indication where the population mean is likely to lie. Point estimation provides estimates which are difficult to interpret, whereas with interval estimation, you can say what level of confidence you have that your population parameter lies within a given interval range. The range is restricted by an upper and a lower limit, and confidence is usually expressed as 95 per cent or 99 per cent. So, if our IQ mean from one sample was 100 and we knew the shape of our sampling distribution, we may be in a position to state that we are 95 per cent confident that the population mean lies between 70 and 130. One point to notice is that the upper and lower limits are equidistant from the sample mean.

You can express the confidence with which you can state that the population parameter of interest will lie between an upper and lower limit, in terms of a percentage or a probability. Obviously, the smaller the interval the closer the sample statistic will be to a true estimate of the population parameter.

One of the images of the researcher presented in the previous chapters is of someone who is seeking empirical evidence for his or her predictions or hypotheses. Such researchers are therefore more interested in testing their hypotheses than in simply making estimations of population parameters from their sample statistics.

In our decision-making process we have to make choices, and in terms of our hypotheses we need to decide between the possibility that there are no effects and the possibility that the effects do exist, as predicted by our research hypothesis. A research hypothesis is referred to as the *alternative hypothesis* (H_1), and the hypothesis of no difference is called the *null hypothesis* (H_0).* Strictly speaking, it is only the null hypothesis which can be tested, and our data analysis indicates whether or not the null hypothesis should be rejected.

When testing hypotheses in a simple experimental situation, we may wish to compare a group which has been subjected to a certain treatment or manipulation (experimental group) with a group which has not received the treatment or manipulation (control group). If

*The symbols H_0 and H_1 are conventions. Some textbooks (e.g. Winer, 1971), however, adopt H_1 and H_2 as representing the null hypothesis and research hypothesis respectively. In addition, we presented Fisher's definition and use of the null hypothesis. Others have disagreed with his particular use of the term and some even with the whole concept of hypothesis-testing (see Heerman and Braskamp, 1970).

we subsequently find that there is a difference between the means of the two groups on our dependent variable, then we may wish to argue that the difference was due to our treatment effect. In order to do this we need to ask the question *'what is the probability that our sample statistic is due to sample variation rather than the different research conditions?'* In more orthodox statistical language, in order to test the null hypothesis we need to assess the relative likelihood of our means having been sampled from the same population or two distinct populations. If two means are sampled from the same population then any difference between them would be simply due to sampling variation, and therefore the null hypothesis should not be rejected. Alternatively, if two means are drawn from two distinct populations then the difference between the means would be regarded as the result of differential treatment conditions. Therefore the null hypothesis should be rejected in favour of the alternative hypothesis.

For example, in Chapter 1 we outlined a possible experiment designed to investigate whether the low productivity rate in a factory was a result of boredom. We suggested that one group of workers might be continually moved between jobs (the experimental group) and their level of productivity compared to that of a group who remained on the same job all day (the control group). Productivity could be measured in terms of the number of units completed by the end of the working day. Let us say that our experimental group produced a mean of 45 units compared to the control's mean of 27 units. The question we must ask ourselves is whether this difference between means was due to our manipulation or simply due to sampling variation (error). If we could produce, as described in sections 3 and 4, a sampling distribution of means (like Figure 5.3), then we could find out the probability associated with the occurrence of *each* mean. This would not, however, provide us with any direct information about the probability of a difference *between* the two means.

In order to obtain this sort of information, it is possible to draw up a distribution which plots the differences between pairs of sample means. Since most means will lie about the centre of the sampling distribution of means, the average difference between them would tend to be small and thus lie around the mean of zero. Therefore, large differences between means would be quite unusual if the samples were drawn from the *same* population. Figure 5.7 is a

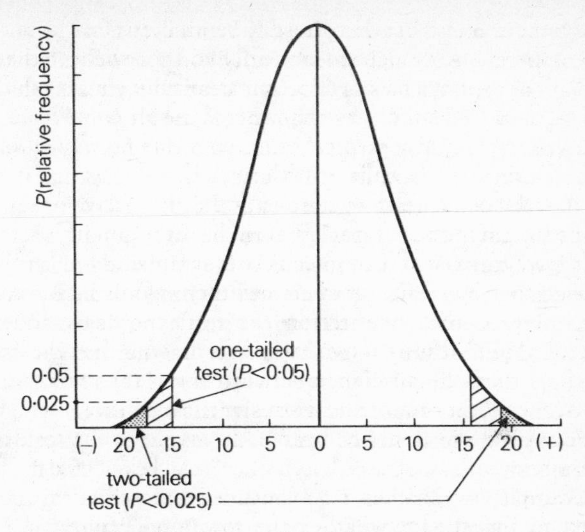

Figure 5.7 A sampling distribution of the difference between means; hatched areas are regions of rejection at $P < 0.05$ and $P < 0.025$ level for a one-tailed or two-tailed test

hypothetical sampling distribution of the differences between means for our factory example.

The difference between our two means in this example was 18 (45 − 27). We can see from the figure that such a difference would be quite rare if our samples were drawn from the same population, since it lies in the 'tail' of the distribution. To find such a difference arising from sampling the same population is so improbable that we could conclude that the difference between our two means is not due to sampling error. Instead, the means are likely to have been sampled from *different* populations, as a result of our experimental manipulation. In effect, we have rejected the null hypothesis, as epitomized by the sampling distribution, and hence found supporting evidence for our alternative, or research, hypothesis.

Several important points arise from the above discussion. First, as already stated, the statistical procedure outlined above allows us to test the null hypothesis only, not a research hypothesis. It is, therefore, not possible to have *a research hypothesis of 'no difference'*. In other words, you cannot design experiments with the aim of testing the null hypothesis directly, by predicting an absence of treatment effects. Although theoretically you may believe that you can explain or predict a lack of effects, statistically your result could have been produced simply by chance. The result could have been produced in the way you predicted, but you have no way of telling. However, despite the fact that it is the null hypothesis that is being tested, it is the convention to state only the alternative or research hypothesis, with the null hypothesis remaining implicitly assumed.

Second, we can assess the probability of mistakenly rejecting the null hypothesis by reference to an arbitrary *significance level*. By convention the minimum criterion for significance is usually set at $P = 0{\cdot}05$. For your results to be statistically significant, the associated probability would need to be equal to or less than this criterion (i.e. $P < 0{\cdot}05$). Other more stringent significance levels which are commonly used are $P < 0{\cdot}01$ and $P < 0{\cdot}001$. If your results are statistically significant, the null hypothesis can be rejected in favour of the alternative hypothesis. This can be represented by an area under the curve called the *region of rejection* (see Figure 5.7). If you do reject the null hypothesis, when in fact it was true, you have made what is called a *Type I error*. The probability of making this type of error is represented by the Greek letter alpha (α), which is the significance level. However, there is also a possibility that you may not reject the null hypothesis when in fact your research hypothesis was true. This is called a *Type II error* and its associated probability is represented by the Greek letter beta (β) (see Table 5.1).

Table 5.1. A decision matrix for accepting or rejecting the null hypothesis (H_0) and the errors associated with making the wrong decisions (Type I and Type II).

		FINDINGS (H_1)	
		Significant	Non-significant
H_0	True	Reject incorrectly Type I (α)	Accept correctly
	False	Reject correctly* ($1 - \beta$)	Accept incorrectly Type II (β)

* Note that the 'power' of the test can be expressed as $1 - \beta$ (see Chapter 3).

There is a third important point about hypothesis-testing. This relates to the fact that a hypothesis can be either directional, and thus able to predict into which tail of the distribution our sample statistic will fall, or non-directional, thus allowing for the possibility of it falling into either tail. A directional hypothesis would therefore be one which states the expected direction of the difference between the group means (e.g. 'female comprehensive school children are more intelligent than male comprehensive school children'). A non-directional hypothesis, on the other hand, would not state any expectation regarding the direction of the difference in means (e.g. 'sex differences exist in the intelligence of comprehensive school children'). The type of hypothesis will have certain consequences for the way in which we test it. *If it is a directional hypothesis then we need a one-tailed statistical test, and if it is a non-directional hypothesis then we need a two-tailed statistical test.* This difference is important because the significance level represents a *total* probability or the total area of rejection. In a two-tailed test this total area must be divided equally between the two tails, which in effect halves the associated probability for each tail. The total probability thus still remains constant (e.g. $P < 0{\cdot}05$) but is shared between the two tails, so that the probability of your statistic falling into any one tail is now halved (e.g. $P < 0{\cdot}025$ – see Figure 5.7).

The Assumptions Underlying Statistical Tests

We will now turn our attention to the general nature and use of statistical tests. However, before embarking on this discussion we need to ask ourselves why statistical tests are necessary at all. After all, if the normal distribution is a useful sampling distribution then surely all our decision-making could be based on this distribution. Unfortunately, the normal distribution is not always the best representation of a sampling distribution. As explained earlier, if we wish to test the significance of the difference between our sample means we require a *sampling distribution of differences between means.* Although for large samples this is likely to approximate to the normal distribution, for a sample size of less than 100 it has been found not to do so, because our estimate of the population parameters is less accurate. In such cases we may need to refer instead to an alternative distribution (e.g. the *t*- or *F*-distribution). More correctly, we would refer to a particular family of distributions to help make a decision between our two hypotheses. We refer to a family of distributions, rather than a single distribution, because the shape of the distribution changes with increasing sample size (N) or degrees of freedom ($N - 1$). Naturally, all these distributions do approximate to the normal as sample sizes increase, owing to the central limit theorem.

Consequently, there are many important distributions which will help us to decide whether or not to reject the null hypothesis. These distributions require you to calculate a test statistic before the appropriate probability can be determined. For example, before you can refer to the t-distribution you need to use a statistical test (i.e. the t-test) in order to calculate a value for t. The probability of occurrence of the particular t-value can then be determined from its sampling distribution. Thus, given that you have calculated a test statistic and that you know the sample size or the number of degrees of freedom, there are tables available which will tell you the critical value of the test statistic at varying levels of probability (usually 0·05, 0·025, 0·01 and 0·001). The Appendix has examples of such tables.

Unfortunately, we are restricted in our use of certain kinds of sampling distributions, and hence particular categories of statistical test, because they can be reliably used only if certain assumptions are met. In other words, when using them we can feel confident that our statistical decision is correct only in situations where the underlying assumptions are met. Without such confidence we may increase the risk of making Type I or Type II errors.

As you may recall from Chapter 3, those tests which are concerned with parameters and require the adherence to certain assumptions are called *parametric tests.* We can improve the accuracy of the estimates of parameters by sampling randomly and by assigning people to groups randomly, thus keeping our observations independent. As we have seen earlier in the chapter, random sampling helps to ensure the *independence of our samples.* In addition, random assignment helps to ensure the *independence of our treatments.* It is not surprising then that independence of observations is one of the assumptions underlying the use of parametric tests. In Chapter 3, we introduced one of the other restrictions or assumptions in that parametric tests usually require that the *data is measured on either an interval or a ratio scale.* The final two assumptions are concerned with the determination of the appropriate sampling distribution. In the present chapter, we have mentioned that the normal distribution is useful for determining sampling distributions through the application of certain theorems, such as the central limit theorem. Thus, if we can assume that the *sample is drawn from a normally distributed population*, then those useful characteristics associated with the normal distribution can be used to calculate our sampling distribution. Also if we can assume that *the samples contain an equal amount of variance* (i.e. the range of scores in each sample is about equal) then we can fulfil a further demand of using parametric tests. The reasoning behind this assumption will be explored more fully in Chapter 9.

Parametric tests, therefore, have a number of general assumptions associated with their use. Consequently, if you suspect that any of these assumptions have been breached, then you *may* need to consider using a 'distribution-free' or non-parametric test. This does not imply that non-parametric tests have no associated assumptions; however, they do generally make less stringent demands (see Chapter 10).

Although some specific forms of statistical test (e.g. analysis of variance) may make additional assumptions, the four main assumptions underlying parametric tests can be summarized as:

(*a*) the population distribution from which the sample was drawn should be normally distributed;

(*b*) the observations should be independent (usually assured by randomly sampling and assigning);

(*c*) the measurements should have been made on either an interval or a ratio scale;

(*d*) there should be homogeneity of variance (i.e. the variances of the populations are equal as reflected in the similarity of the sample variances).

The assumptions underlying parametric tests can, if strictly adhered to, heavily constrain the nature of our investigations. However, these are theoretical assumptions, and they are not always kept in practice. The importance of these assumptions can be small or great according to the particular circumstances of your investigation. There are many researchers who argue that the assumptions should be regarded as guidelines only and that there are very few situations where the use of parametric statistics results in dire consequences.

They imply that most parametric tests are sufficiently *robust* not to be seriously affected by the violation of some of their assumptions. Perhaps the two less critical assumptions are homogeneity of variance, and the normality of the population's distribution.

Although we can test for homogeneity of variance (Hartley, 1950), such tests are extremely sensitive, and may lead the researcher to reject the use of a parametric test too quickly. Lindquist (1953) has demonstrated that parametric tests are sufficiently robust to deal with relatively large differences in sample variances as long as *the sample sizes are equal*. In the case of unequal sample sizes, it is wise to test for homogeneity of variance and such a test can be found in Winer (1971). The lack of any firm evidence for a normally distributed population also appears to have little consequence for the sensitivity and usefulness of most parametric tests. Nevertheless, you should still be cautious when interpreting the results of parametric tests used in situations where your sample size may be too small (e.g. $N < 5$) to provide any reliable indications of the shape of the population distribution.

There is also some limited flexibility with regard to the assumption of level of measurement scale, although knowing what is truly legitimate with respect to your own investigation is not easy and care must be taken (see Chapter 3). For those situations where the breaching of this assumption may not militate against the use of a parametric test see Chapter 8. On some occasions it is possible to normalize ordinal data. Fisher and Yates have developed a procedure for transforming an individual's rank data into normal scores (see Winer, 1971).

In general, we recommend that if a number of the assumptions associated with using parametric tests appear to be breached, and an appropriate non-parametric test is available, then the parametric test should not be used. (See Chapter 3 for more detailed discussion about the choice of appropriate tests.) If, however, no suitable non-parametric test is available, then you should consider seeking expert advice. For example, it may be possible to transform your data (e.g. an arc sine transformation) in order to remove both heterogeneity of variance and non-normality.

The reason why psychologists search for various means of overcoming or disregarding the restrictions imposed by parametric tests is that there are disadvantages associated with the use of non-parametric tests:

(*a*) *Non-parametric tests cannot readily be used to make statements about the differences between means*, because they compare distributions rather than parameters of distributions.

(*b*) *Non-parametric tests in general cannot be used for investigations involving more than two independent variables*. If you are comparing levels within one variable then there is only a limited disadvantage in using a non-parametric test.

(*c*) *Very few non-parametric tests are capable of dealing directly with the more complex experimental designs*. This is because more complex designs produce interactional data, which can be of great interest to the researcher but for which there are few readily accessible non-parametric tests available (see Chapter 10).

(*d*) *Non-parametric tests are generally less powerful than their parametric equivalents* (see Chapter 3). However, if the assumptions associated with parametric tests are not respected then there may be an increased probability of making a Type I error (i.e. accepting that your findings are significant when they in fact are not). Equally, if you use a less powerful test than you could have done, then you may be increasing the probability of making a Type II error (i.e. accepting that your findings are not significant when in fact they are).

You should try to use the most powerful test at your disposal. It is therefore useful to know the *power efficiency* of the various tests. This allows you to compare the power of a non-parametric test with its equivalent parametric test. Power efficiency is expressed as *the ratio of the size of the samples required to make the two tests equal in power*. For example, for a non-parametric test to have the equivalent power of its parametric counterpart it may require a sample size 25 per cent greater than that required by the parametric test. Its power efficiency can therefore be expressed as 75 per cent of its equivalent parametric test. In fact, the power efficiency of most non-parametric tests lies in the 90 per cent region.

Summary

This chapter explored the important concepts and assumptions underlying statistical tests. This involved a discussion of the best ways of drawing samples which are representative of populations, and the method by which this representativeness can be assessed statistically. A distinction was made between drawing samples in order to make estimations of population parameters, and drawing samples in order to test hypotheses. 'Hypothesis-testing' was presented as essentially a statistical decision-making process, and one of the major uses of statistical tests. Finally, the general reasoning behind the use of statistical tests was discussed and the advantages and disadvantages of certain categories of test outlined.

6 Measures of Association and Correlation

This chapter looks at ways of measuring association between pairs of nominal-, ordinal- or interval-level measured variables. For pairs of nominal-level variables we advocate a method of measuring association strength over a measure that simply detects whether an association is present or not. For pairs of interval-level variables we introduce a method that measures the strength of correlation. In addition, we describe a method of estimating the population correlation coefficient. We also show that correlation coefficients for independent samples can be compared. We point out that ordinal-level data can be correlated using the same procedure as for interval-level data. Finally, we discuss the problems inherent in any attempt to interpret measures of association or correlation.

Nominal Association

In this section we will show the inadequacy of tests of independence, such as chi-square, and we will advocate an alternative, a measure of association strength. Chi-square is a test of association between two nominal or categorical variables. For instance, you could use chi-square to see if there is an association between students passing or failing a particular examination after having been taught by one of three different methods. Recall that chi-square should only be used if each observation in the contingency table is *independent* of every other observation in the table. In this particular case, each student *either* passed or failed and was taught by only *one* of the three teaching methods.

On the face of it, chi-square tests are amongst the easiest to apply and can be used on almost any social science data. For instance, test scores that consist of interval-level data can be easily reduced to categorical data, e.g. those who scored more than the pass mark and those who scored less than the pass mark. Similarly ordinal-level or ranked scores can be categorized into those who passed and those who did not. However, such transformations involve loss of information and so should be avoided when other tests can be appropriately used.

An often-quoted and generally agreed rule of thumb that should be followed when you apply chi-square is that none of the expected frequencies in the cells of the contingency table should fall below five.

It may have occurred to you that one way to achieve higher expected values in the cells of a contingency table is to 'collapse' or 'pool' categories *once you have seen the data from your study*. For instance, the contingency table containing these expected values:

	Methods of teaching		
	A	B	C
pass	50	47	56
fail	4	14	8

could be collapsed to:

	Methods of teaching	
	A or B	C
pass	97	56
fail	18	8

to overcome the 'problem' of the cell containing an expected value of 4.

However, *this procedure should be undertaken with the utmost caution*. Recall that the rationale underlying chi-square assumes that your sample is random and that the categories into which your observations *may* fall have been chosen by you *in advance*. Obviously the way in which you pool your categories affects the inferences you can draw because you have altered the randomness of your sample in a *deliberate* way. To our mind it is better to avoid such manipulations when they are not based on an adequate theoretical foundation, as we saw in Chapters 3 and 5.

This example of an effort to achieve a significant result is somewhat analogous to another problem with the application of chi-square. Recall that chi-square tests to see if there is an association, or alternatively independence, between two nominal-level variables. It follows that if your sample size, *N*, is very large (as it should be for best application of the test), then any small degree of association between the variables will produce a statistically significant result. Consider,

	Methods of teaching	
	A	B
pass	2000	1900
fail	1900	2000

Following the formulae for chi-square: $\chi^2 = 5{\cdot}03$ with 1 DF. This result is significant at the 5 per cent level: in other words, there is significant evidence for an association between method of instruction and relative proportion of passes. But look again at the similarity of the numbers in the contingency table!

It is clear that chi-square detects virtually any deviation from a strict independence between two variables when *N* is large. Given a significant result, it is of course possible to draw the inference stated in the preceding paragraph. But surely most variables are in some degree related. It follows that, by choosing a suitably large value of *N*, a non-naive social scientist could produce evidence of a significant association between any two nominal variables. Similarly, if this non-naive social scientist found a nearly significant chi-square on first attempt he could collect some more data and then re-do the chi-square test and find 'what he was looking for'! As we saw in Chapter 3, given a large enough sample size, the chances are very good that you can demonstrate an 'expected' significant result.

With regard to chi-square, this problem has led to the development of tests which measure nominal-level *association strength*, in contrast to detecting association or independence. The result of the test to be discussed next is not crucially dependent on sample size. It gives an indication of how your knowledge of the way in which people have been categorized on one variable can help predict their categorization on another. This measure of *predictive association* can allow you to draw inferences about the importance of a relationship in addition to saying if a relationship is present or not.

Goodman–Kruskal Index of Predictive Association (Lambda)

Suppose that for a sample of the general population the joint probability distribution of two variables, social class and educational qualifications, is as follows:

	'Upper class'	'Middle class'	'Working class'	
No qualifications beyond secondary school level	0·10	0·15	0·30	0·55
Some qualifications beyond secondary school level	0·05	0·30	0·10	0·45
	0·15	0·45	0·40	1·00

In other words, the probability of a person picked at random from the sample being a member of the 'upper class' and having no qualifications beyond secondary school level is 0·10 or 10 per cent, the probability of a person being a member of the 'middle class' and having some qualifications beyond secondary school level is 0·30 or 30 per cent, and so on. The whole probability distribution adds up to 1·00 or 100 per cent because each of the people in the sample *must* fall into *one* of the cells of the contingency table.

Now if you knew absolutely nothing about which social class a person belongs to, what level of qualifications would you bet he or she has? Obviously the answer is 'no post-secondary qualifications', because 0·55 or 55 per cent of the sample are categorized as having none whereas only 0·45 or 45 per cent are categorized as having some.

However, if you knew which social class category a person belonged to this could give you more information about his or her qualifications. For instance, if a person was categorized as 'working class' he or she is three times more likely to have no advanced qualifications as to have some. This shows that there is a *predictive association* between variables. Knowing a person's categorization on one variable tells us something (although nothing definite) about the likelihood of him or her being categorized on a particular category of the second variable. This idea is the basis for Goodman and Kruskal's index of predictive association, known more simply as *lambda*. Lambda gives us a measure of the probability of error in making a prediction of a person's categorization on one nominal variable, given that we know his or her categorization on the other nominal variable. Lambda values can vary between 0 and 1: 0 means that there is no reduction in the probability of making an erroneous prediction, while 1 means that probability of error has been reduced to zero and so the probability of a successful prediction is certainty. Of course, in most real-life studies the computed lambda value lies between these two extremes.*

* Here we have discussed and illustrated the symmetric version of lambda, called λ_{ab}. Other versions of lambda are given in Hays (1963, p. 606).

Step-by-Step Procedure

Goodman–Kruskal index of predictive association

Step 1	Draw up the contingency table.
Step 2	Find the row totals, column totals and grand total.
Step 3	Find the highest score in each of the columns. Find the highest score in each of the rows.
Step 4	Find the highest column total. Find the highest row total.
Step 5	Note the grand total.
Step 6	Add the highest scores for each of the columns to the highest scores for each of the rows, i.e. sum the scores obtained in step 3.
Step 7	Add the highest column total to the highest row total, i.e. sum the scores obtained in step 4.
Step 8	Subtract step 7 from step 6.

Worked Example

Goodman–Kruskal index of predictive association

In studying the possible relationship between smoking and cancer, the data below were obtained. Of 1,425 subjects investigated 695 were smokers and 730 were non-smokers. Of the smokers, 230 had developed lung cancer by a certain age, 465 had not. Of the non-smokers, 78 had developed lung cancer and 652 had not. What is the strength of association between smoking and lung cancer?

Step 1

	Smokers	Non-smokers
cancer	230	78
no cancer	465	652

Step 2

230	78	308
465	652	1117
695	730	1425

Step 3 Highest in first column is 465.
Highest in second column is 652.
Highest in first row is 230.
Highest in second row is 652.

Step 4 Highest total for a column is 730.
Highest total for a row is 1117.

Step 5 Grand total is 1425.

Step 6 Sum of step 3 is $465 + 652 + 230 + 652 = 1999$.

Step 7 Sum of step 4 is $730 + 1117 = 1847$.

Step 8 Step 6 − step 7 = 152.

Step 9 Multiply step 5 by two.

Step 10 Subtract step 7 from step 9.

Step 11 Divide step 8 by step 10.

Step 12 Translate the result of the test back in terms of your study.

Interval-level Correlation

The scores of individuals on two interval-level variables can be represented in the form of a *scattergram*. For example, the relationship between height and weight might appear as in Figure 6.1. Each point in the scattergram represents an individual person. As increases in height are usually accompanied by increases in weight you would perhaps expect evidence of a positive relationship between the two variables, although not a perfect correlation because there are short fat and tall thin individuals! For continuous or interval-level variables such as weight, age and most scores on psychological tests (for a full discussion see Chapters 3 and 8), we can compute an interval-level correlation coefficient that makes the

Step 9 $2 \times$ step 5 $= 2 \times 1425 = 2850$.

Step 10 Step 9 − step 7 $= 2850 - 1847 = 1003$.

Step 11 $\frac{\text{Step 8}}{\text{Step 10}} = \frac{152}{1003} = \underline{\underline{0{\cdot}15}}$.

Step 12 This says that if we know *either* whether a person was a smoker or not *or* whether a person had developed lung cancer or not, this knowledge helps us predict a person's categorization on the other variable in that the probability of making an erroneous prediction has been reduced by about 15 per cent.* However, this probability is nowhere near 100 per cent and so knowing a person's categorization in one of the variables has not helped us a great deal, in this instance. Notice that in this example N is very large, in fact 1425, so on the basis of our previous discussions you might consider it possible that a chi-square analysis would have produced evidence of a significant association. Re-analyse the data using chi-square; alternatively, refer to Colin Robson's *Experiment Design and Statistics in Psychology*, pp. 94–7.

* It is possible to test the significance of an obtained lambda but the computations are complicated. The interested reader is referred to Goodman and Kruskal (1954).

---

most use of the level of data available. The most commonly used is *Pearson's product-moment correlation coefficient*, sometimes called 'Pearson's r' or just 'r'.

Product-moment Correlation Coefficient (r)

The correlation coefficient produced by the formula used in calculating the product-moment correlation coefficient will range from −1 to +1, where a value of +1 describes a perfect positive relation in that all points lie along a straight line and as the scores on the X-axis *increase* those on the Y-axis *always increase* (Figure 6.2*a*).

A value of −1 describes a perfect negative relation. All points lie

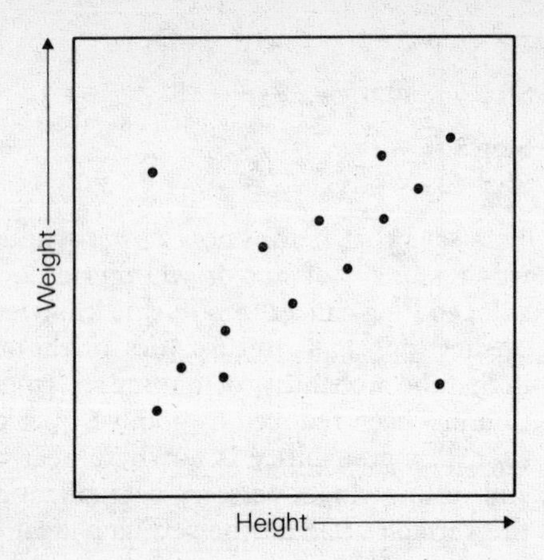

Figure 6.1

along a straight line and as scores on the X-axis *decrease* those on the Y-axis *always increase* (Figure 6.2*b*). Most *observed* relationships between pairs of scores are found between the two extremes of a perfect positive or a perfect negative relationship!

Note that the words 'always increase' or 'always decrease' are used in definitions of perfect correlations. The slope of the line connecting the points on the scattergraph is immaterial. However, one individual who does not follow the ordering implied in the definition will reduce the perfect correlation coefficient (Figure 6.3).

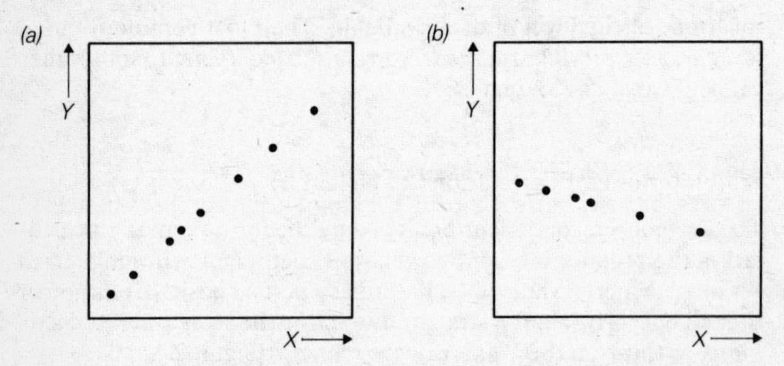

Figure 6.2 (*a*) a perfect positive correlation, (*b*) a perfect negative correlation

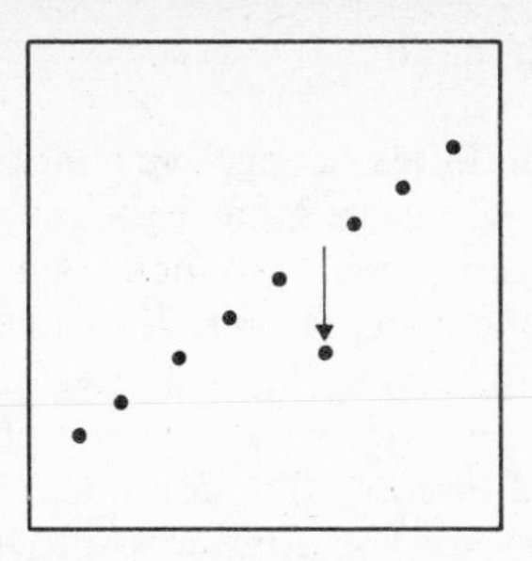

Figure 6.3

The product-moment correlation coefficient is not concerned with the magnitude of scores on each of the two variables. If it was appropriate to compare the population means of the variables represented on the X- and Y-axes and they were shown to be similar or significantly different (by a related t-test), it is of no consequence in the computation of their degree of correlation.

One main assumption underlying application of the product-moment correlation coefficient is that the sample distributions, sometimes called the *marginal distributions*, for each of the two variables to be correlated are similar. If the ranges of X and Y are very different then any obtained correlation will be restricted, in that it will not have the freedom to range between $+1$ and -1.

What is the rationale underlying computation of the product-moment correlation coefficient? Basically, the correlation coefficient quantifies the degree of linear relationship between two variables. This is best illustrated if we consider the formula for the correlation coefficient in deviation score form:

$$r = \frac{\Sigma xy}{\sqrt{(\Sigma x^2)(\Sigma y^2)}}$$

Here $x = X - \bar{X}$ where X is a raw score and $\bar{X}$ is the mean of the X-scores, and y is similarly related to the Y-scores. When there is a positive relationship between the x- and y-scores, high x-scores coupled with high y-scores will produce positive deviations and result in positive cross-products; low x-scores coupled with low y-scores will also result in positive cross-products, since the product of two negative deviations must be positive. The sum of the cross-products (Σxy) will thus be positive and the size of its sum will increase in line with the strength of the *positive* correlation between X and Y scores.

Conversely, when there is a negative relationship between the X- and Y-scores, high x-scores will tend to be coupled with *low* y-scores and so produce a negative cross-product. High y-scores coupled with low x-scores will, of course, also produce a negative cross-product. The sum of cross-products will thus be negative and the size of the sum will increase in line with the strength of the *negative* correlation between X- and Y-scores. The denominator of the formula for r simply restricts the size of positive cross-products to 1 and negative cross-products to -1.

Before we move on to the computation of r, let us re-emphasize that the product-moment correlation coefficient is a test for a linear (straight-line) relationship between two variables. Non-linear relationships will not be detected. Small values of r suggest *either* that there is no linear relationship between variables *or* that the type of relationship is non-linear. If you suspect, or anticipate, a non-linear relationship between your two variables, draw a scattergram *before* doing any calculations. Methods of evaluating non-linear relationships are beyond the scope of this book but may be found in more advanced statistical texts such as Hays (1963). Figure 6.4 gives one example of one possible non-linear relationship between level of arousal and examination performance. Both those who are under-

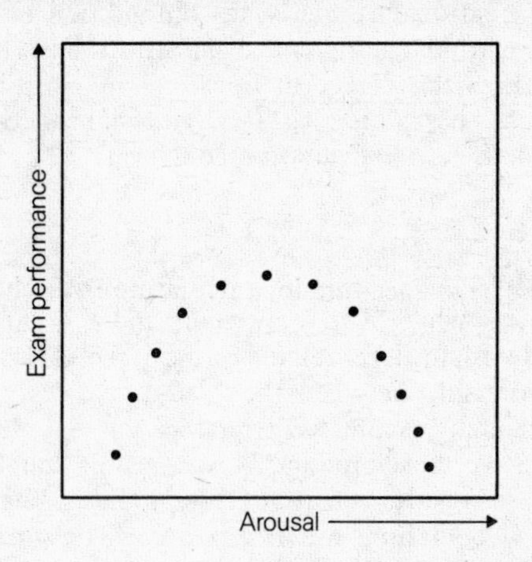

Figure 6.4

aroused (sleepy) and those who are over-aroused (very nervous) do badly in exams, whilst those who take it in their stride do best!

Another problem with a mechanical or unthinking computation of correlation coefficients is that the presence of 'outliers' can help to produce a positive or negative r value from data that in all other respects is virtually uncorrelated. For example, the data represented in Figure 6.5 would produce a positive r; but perhaps the two people producing the outlier scores were not English and had difficulty in understanding the questionnaires! Methods of *exploratory data analysis* facilitate a thorough examination of data and will highlight problems such as these. For more detail see Tukey (1977) and Hartwig and Dearing (1979).

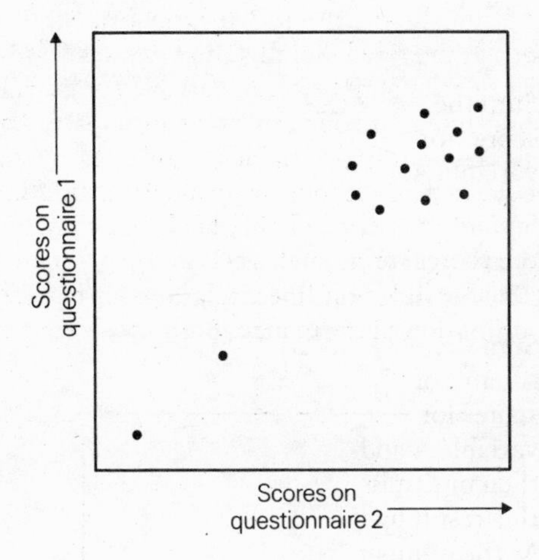

Figure 6.5

The step-by-step procedure for calculating r which follows does not use deviation scores. It uses raw scores, for ease of computation.

Step-by-Step Procedure

Product-moment correlation coefficient

	Calculations	**Usual symbols**
Step 1	Lay out the scores for the two variables in pairs.	
Step A2	Sum the scores for variable *A*.	ΣA
Step A3	Square each of the scores for variable *A*.	
Step A4	Sum the squares of scores for variable *A* and then multiply this result by *N*, the number of pairs of scores.	$N\Sigma A^2$
Step B2–4	Repeat the above three steps for the *B* scores.	ΣB, $N\Sigma B^2$
Step A5	Square the result of step A2.	$(\Sigma A)^2$

Worked Example

Product-moment correlation coefficient

Scores on two intelligence tests, Test A and Test B, were obtained for each of ten students. Do people who score high on Test A also score high on Test B, and do people who score low on Test A also score low on Test B? Or do opposite or no such tendencies exist? In other words, are the scores on the two tests correlated?

Step 1

Student	1	2	3	4	5	6	7	8	9	10
Test A	15	19	13	12	6	30	22	8	1	15
Test B	96	50	40	30	10	100	10	12	5	30

Step A2 $\Sigma A = 141$

Step A3 225 361 169 144 36 900 484 64 1 225

Step A4 $\Sigma A^2 = 2609, N\Sigma A^2 = 10 \times 2609 = 26090$

Step B2 $\Sigma B = 383$

Step B3 9216 2500 1600 900 100 10000 100 144 25 900

Step B4 $\Sigma B^2 = 25485, N\Sigma B^2 = 10 \times 25485 = 254850$

Step A5 $(\Sigma A)^2 = 19881$

Step A6	Subtract step A5 from step A4.	$N\Sigma A^2 - (\Sigma A)^2$
Step B5–6	Repeat the above two steps for the *B* scores.	$(\Sigma B)^2, N\Sigma B^2 - (\Sigma B)^2$
Step 7	Multiply step A2 by step B2.	$\Sigma A \Sigma B$
Step 8	Multiply each *A* score by its paired *B* score and sum this result over the whole set of pairs. Then multiply this result by *N*.	$N\Sigma AB$
Step 9	Subtract step 7 from step 8.	$N\Sigma AB - \Sigma A \Sigma B$
Step 10	Take the square root of the result of step A6.	$\sqrt{N\Sigma A^2 - (\Sigma A)^2}$
Step 11	Take the square root of step B6.	$\sqrt{N\Sigma B^2 - (\Sigma B)^2}$
Step 12	Multiply step 10 by step 11.	$\sqrt{[N\Sigma A^2 - (\Sigma A)^2]}\sqrt{[N\Sigma B^2 - (\Sigma B)^2]}$
Step 13	Divide step 9 by step 12.	$r = \dfrac{N\Sigma AB - \Sigma A \Sigma B}{\sqrt{[N\Sigma A^2 - (\Sigma A)^2]}\sqrt{[N\Sigma B^2 - (\Sigma B)^2]}}$
Step 14	Translate the result of the test back in terms of your study.	

Step A6 $N\Sigma A^2 - (\Sigma A)^2 = 26090 - 19881 = 6209$

Step B5 $(\Sigma B)^2 = 146689$

Step B6 $N\Sigma B^2 - (\Sigma B)^2 = 254850 - 146689 = 108161$

Step 7 $\Sigma A \times \Sigma B = 141 \times 383 = 54003$

Step 8 $\Sigma AB = 1440 + 950 + 520 + 360 + 60 + 3000 + 220 + 96 + 5 + 450 = 7101$
$N\Sigma AB = 10 \times 7101 = 71010$

Step 9 $N\Sigma AB - \Sigma A\Sigma B = 71010 - 54003 = 17007$

Step 10 $\sqrt{N\Sigma A^2 - (\Sigma A)^2} = 78{\cdot}80$

Step 11 $\sqrt{N\Sigma B^2 - (\Sigma B)^2} = 328{\cdot}88$

Step 12 $78{\cdot}80 \times 328{\cdot}88 = 25915{\cdot}74$

Step 13 $\frac{17007}{25915{\cdot}74} = \underline{\underline{0{\cdot}656}}$

Step 14 People who score high on Test A also tend to score high on Test B and people who score low on Test A also score low on Test B. However, this positive correlation is not perfect.

Significance of a Correlation Coefficient

Whether a computed sample correlation coefficient suggests that the correlation in the population is significantly different from zero correlation or not is crucially dependent on sample size, i.e. number of pairs of scores. The larger the sample size the lower the obtained or computed coefficient can be and still be significant. Table II in the Appendix presents critical values of obtained correlation coefficients that are significant at the 0·05 (5 per cent) level of significance. If an obtained correlation is equal to or greater than the table value for N pairs of scores, then your obtained correlation is significantly different from zero. In this case 0·656 is significant at the 5 per cent level, for a two-tailed test. Notice that an obtained correlation of 0·656 would be significant at the 5 per cent level if it was based on a sample of 10 people but *no less*.

Step-by-Step Procedure

Confidence intervals for a correlation coefficient

Step 1 Transform r to a z-score, z_r, using Table III in the Appendix.

Step 2 Note sample size (i.e. number of pairs of scores) = N.

Step 3 Compute the standard error of z_r

$$= \sqrt{\frac{1}{N-3}}$$

Step 4 Either:

(*a*) 95 per cent of the area under the normal curve lies within 1·96 standard deviations of the mean so the 95 per cent confidence interval is 1·96 multiplied by the standard error above and below the obtained value of z_r or, in other words:

Confidence Intervals Around an Obtained Correlation Coefficient

In some studies you may not be especially interested if an obtained sample correlation suggests that the correlation in the population is significantly different from a zero correlation. Rather you may be interested to estimate the true population value of r, denoted by the Greek letter ϱ (rho). To do this we compute a confidence interval around our obtained sample value of r. Conventionally we compute the 95 per cent (or 99 per cent) confidence interval.

To compute the confidence interval around an obtained value, r is transformed to a z-score because difficulties resulting from the skewness of the sampling distribution of the correlation coefficient make direct interpretation difficult. In Chapter 5 we discussed z-scores and the usefulness of the standard normal distribution. The distribution of z_r (i.e. z-scores of correlation coefficients) is approximately normal and has the same variability for a given sample size no matter what the true value of ϱ, and so the z_r transformation may be used to obtain the confidence limits for your obtained correlation, r.

Worked Example

Confidence intervals for a correlation coefficient

We will use the value 0·656 obtained earlier for the correlation between scores in the two intelligence tests.

Step 1 $r = 0{\cdot}656$; from Table III, $z_r = 0{\cdot}786$.

Step 2 $N = 10$.

Step 3 Standard error of $z_r = \sqrt{\dfrac{1}{10 - 3}} = 0{\cdot}378$.

Step 4 (*a*) 95 per cent confidence interval

$$z_r \pm 1{\cdot}96 \times \sqrt{\frac{1}{N-3}}$$

Or:

(*b*) 99 per cent of the area under the normal curve lies within 2·57 standard deviations of the mean, so the 99 per cent confidence interval around your obtained z_r is 2·57 multiplied by the standard error above and below your obtained value of z_r or, in other words:

$$z_r \pm 2{\cdot}57 \times \sqrt{\frac{1}{N-3}}$$

Step 5a For the 95 per cent confidence interval first compute

step 1 + 1·96 × (step 3)

which is the top limit on the confidence interval; then compute

step 1 − 1·96 × (step 3)

which is the bottom limit on the confidence interval.

Step 5b Repeat step 5*a* but use 2·57 instead of 1·96.

Steps 6a and b Using Table III, convert the top *and* bottom limits of your confidence intervals (which are expressed in *z*-scores) back to correlation coefficients.

Steps 7a and b You are now in a position to say that you are 95 per cent (or 99 per cent) sure that the true population value of the correlation between the two variables lies between the top and bottom limits. These limits, of course, also contain your obtained sample correlation coefficient.

$z_r \pm 1{\cdot}96 \times 0{\cdot}378$

(*b*) 99 per cent confidence interval

$z_r \pm 2{\cdot}57 \times 0{\cdot}378$

Step 5a Top limit of 95 per cent confidence interval
$= 0{\cdot}786 + (1{\cdot}96 \times 0{\cdot}378) = 1{\cdot}527$

Bottom limit of 95 per cent confidence interval
$0{\cdot}786 - (1{\cdot}96 \times 0{\cdot}378) = 0{\cdot}046$

Step 6a z-score of $1{\cdot}526 = 0{\cdot}910$ correlation.
z-score of $0{\cdot}046 = 0{\cdot}046$ correlation.

Step 7a You are 95 per cent sure that the true population value of the correlation between the two variables lies between 0·046 and 0·910. In graphic terms:

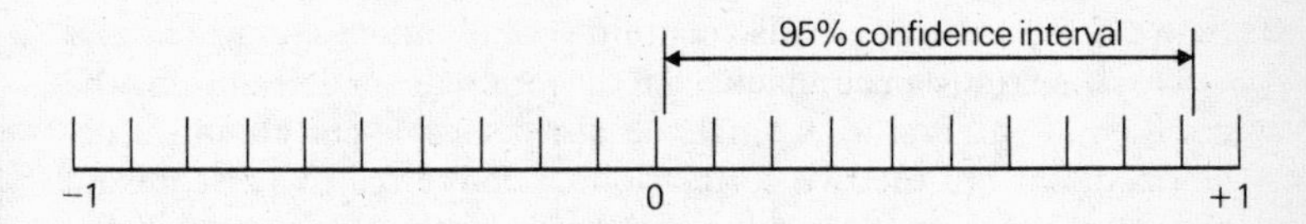

Notice that this 95 per cent confidence interval *does not* include a correlation of 0.

The Difference Between Two Correlation Coefficients for Independent Samples

Suppose you administered two tests of intelligence, one numeric and the other verbal, to both male and female college students. You find that the correlation between the verbal and numeric measures is 0·75 within your male sample ($N_1 = 50$) whilst in your female sample ($N_2 = 60$) the correlation is 0·5. Is verbal and numeric intelligence more closely related in males than females? Or, in other words, is the obtained correlation of 0·75, r_1, significantly higher than 0·5, r_2?

This question can be answered if the rationale for determining the confidence interval for a single obtained sample correlation is developed. Each of the two correlations is converted to a z-score and the standard error of the difference between the two values of z_r is given by:

$$S(z_{r_1} - z_{r_2}) = \sqrt{S^2_{z_{r_1}} + S^2_{z_{r_2}}} = \sqrt{\frac{1}{N_1 - 3} + \frac{1}{N_2 - 3}}$$

Step 5b For the 99 per cent confidence interval:

Top limit 0·786 + (2·57 × 0·378) = 1·757
Bottom limit 0·786 − (2·57 × 0·378) = −0·185

Step 6b *z*-score of 1·757 = 0·942 correlation.
z-score of −0·185 = −0·183 correlation.

Step 7b You are 99 per cent sure that the true population value of the correlation between the two variables lies between −0·183 and 0·942. In graphic terms:

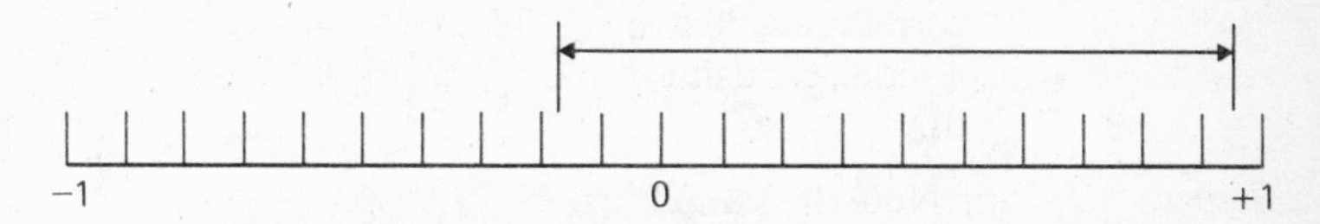

Notice that this 99 per cent confidence interval *does* include a correlation of 0.

If we divide the difference between our two values of z_r by this standard error of the difference, the ratio

$$z = \frac{z_{r_1} - z_{r_2}}{\sqrt{1/(N_1 - 3) + 1/(N_2 - 3)}}$$

can be calculated. Since this ratio is standardized by the *z*-transformation it follows that if the ratio is greater than or equal to 2·58 there is a difference between the two correlations which is significant at the 1 per cent level. If the ratio is greater than or equal to 1·96 the correlations are significantly different at the 5 per cent level.

Step-by-Step Procedure

Difference between two correlation coefficients for independent samples

	Calculations	Usual symbols
Step 1	(*a*) Transform the first correlation, r_1, to a *z*-score, z_{r_1}, using Table III. (*b*) Transform the second correlation, r_2, to a *z*-score, z_{r_2}, using Table III.	
Step 2	(*a*) Note the sample size of the first correlation, N_1. (*b*) Note the sample size of the second correlation, N_2.	
Step 3	Subtract three from the result of step 2(*a*).	$N_1 - 3$
Step 4	Divide one by the result of step 3.	$\frac{1}{N_1 - 3}$
Step 5	Subtract three from the result of step 2(*b*).	$N_2 - 3$
Step 6	Divide one by the result of step 5.	$\frac{1}{N_2 - 3}$
Step 7	Add the result of step 4 and step 6 and find the square root of this sum.	$\sqrt{\frac{1}{N_1 - 3} + \frac{1}{N_2 - 3}}$
Step 8	Subtract the result of step 1(*b*) from the result of step 1(*a*).	$z_{r_1} - z_{r_2}$

Worked Example

Difference between two correlation coefficients for independent samples

Using the correlations of 0·75 and 0·5 from p. 100.

Step 1 (a) $r_1 = 0{\cdot}75$ From Table III $z_{r_1} = 0{\cdot}973$
(b) $r_2 = 0{\cdot}50$ From Table III $z_{r_2} = 0{\cdot}549$

Step 2 (a) $N_1 = 50$
(b) $N_2 = 60$

Step 3 $N_1 - 3 = 50 - 3 = 47$

Step 4 $$\frac{1}{N_1 - 3} = \frac{1}{47} = 0{\cdot}0213$$

Step 5 $N_2 - 3 = 60 - 3 = 57$

Step 6 $$\frac{1}{N_2 - 3} = \frac{1}{57} = 0{\cdot}0175$$

Step 7 $$\sqrt{\frac{1}{N_1 - 3} + \frac{1}{N_2 - 3}} = \sqrt{0{\cdot}0213 + 0{\cdot}0175}$$
$$= 0{\cdot}1970$$

Step 8 $z_{r_1} - z_{r_2} = 0{\cdot}973 - 0{\cdot}549 = 0{\cdot}424$

Step 9 Divide the result of step 8 by the result of step 7.

$$z = \frac{z_{r_1} - z_{r_2}}{\sqrt{\frac{1}{N_1 - 3} + \frac{1}{N_2 - 3}}}$$

Step 10 If the z-value obtained in step 9 is equal to or greater than 2·58 the difference between the two correlations is significant at the 1 per cent level. If the value is equal to or greater than 1·96 the difference between the two correlations is significant at the 5 per cent level.

Ordinal (Ranked) Correlation: Spearman's Rho

Sometimes the scores you wish to correlate will be in the form of ranks. As we saw in Chapter 5, this may be because you have non-parametric data or because your data do not satisfy the parametric assumptions. When this is the case you can simply apply the step-by-step procedure for the product-moment correlation but use ranks instead of raw scores. Turn back to the worked example for the product-moment correlation coefficient given on page 93. Convert scores *within* each test to rankings with the highest score being given a rank of 1 and the lowest score a rank of 10. Your conversion should look like this:

Student	1	2	3	4	5	6	7	8	9	10
Test A	4·5	3	6	7	9	1	2	8	10	4·5
Test B	2	3	4	5·5	8·5	1	8·5	7	10	5·5

Notice that students 1 and 10 share the same score, 15, on Test A and so they also have to share ranks 4 and 5 by averaging: $(4 + 5)/2 = 4·5$. Rank 6 goes to the next highest score. Within Test B ranks have also been averaged between students sharing the same score.

Now, follow the step-by-step instructions and work out the correlation for the ranked scores. You should find it to be 0·651.

When you have used ranks in your calculations the resulting correlation is called *Spearman's rho* rather than Pearson's r.

Step 9
$$\frac{z_{r_1} - z_{r_2}}{\sqrt{\frac{1}{N_1 - 3} + \frac{1}{N_2 - 3}}} = \frac{0{\cdot}424}{0{\cdot}1970} = 2{\cdot}152$$

Step 10 The difference between the two correlations is significant at the 5 per cent level. Numeric and verbal intelligence are more closely related in the male sample than the female sample.

Interpreting Association and Correlation

We have discussed and illustrated the Goodman–Kruskal lambda index of association strength for two nominal-level variables and the Pearson product-moment correlation coefficient for pairs of interval-level scores. We have also seen that the step-by-step procedure for Pearson's r can be used to measure the correlation between pairs of ordinal-level scores.

However, none of the indexes supply information about *why* two variables are associated or related. As we saw in Chapter 4, this question can only be answered by considering the two variables themselves and any theory we have which *predicted* or which can now *explain* the obtained relationship between the variables. Interpretation of obtained correlation or association is the province of the psychologist, *not the statistician*.

For instance, in a study you may have obtained a positive correlation between people's height and their reaction time in pressing a button to indicate that they have seen a message presented via a tachistoscope. Any explanation to describe this correlation may be tenuous to say the least (perhaps greater distance between eye and finger!). Other correlations may be easier to explain, for instance a correlation between degree of extraversion and talkativeness.

Perhaps the major danger in interpretation of obtained association or correlation lies in a tendency to infer *causes* – in other words, to say a person's score on variable *A causes* his score on variable *B*. For instance, you might find that degree of smoking (e.g. number of cigarettes smoked a day) is *correlated* with shorter lifespan. But we cannot say from the results of a correlational analysis that increased smoking *causes* earlier death. It may be that smoking is just an indication of a stressful lifestyle where people smoke in an attempt to relieve stress. Predisposition to stress may cause heart disease which results in an earlier death. In fact, as we saw in Chapter 4, an *experiment* provides the best conditions for inferring causality. But perhaps you may not think it ethical to force a random sample of non-smokers to smoke forty cigarettes a day! In this case it may be possible to use methods of partial correlation to 'partial-out' the effects of predisposition to stress, measured by a questionnaire, and see if a correlation between degree of smoking and lifespan vanishes. Methods like these are elaborated in Chapter 11.

One problem with correlation that you will often face, but which we have not yet discussed, is the problem of interpreting large correlation matrices. For example, suppose you gave three measures of creative ability to a sample of 50 subjects. You predict that scores on the measures will be positively correlated, and you obtain the intercorrelations shown below.

	Test 1	Test 2	Test 3
Test 1	1·00	0·86	0·78
Test 2	0·86	1·00	0·65
Test 3	0·78	0·65	1·00

This pattern of high positive intercorrelations indeed suggests that someone who scores high on one test will also score high on the other two. Conversely, someone who does badly on one test will also tend to do badly on the remaining two tests. Intercorrelations between more than two measures laid out in this way are called an *intercorrelation matrix*. Notice that the correlation of scores on a test with the *same* scores results in a correlation of 1, the diagonal of the matrix. Correlation analyses performed by computer are often presented in this redundant way.

Suppose now that you gave ten personality/cognitive tests to a sample of 50 subjects and as a result you found only two correlations significant at the 5 per cent level out of the possible 45. Suppose that you expected, on theoretical grounds, many more. How would you interpret this result? Recall the rationale of testing the significance of a sample correlation coefficient. If it is significant at the 5 per cent level this means there is a 5 per cent chance that the correlation in the population is zero – in other words, there is a 5 in 100 chance of a spuriously significant correlation. In this hypothetical study two significant correlations were obtained out of a possible 45, just under 5 per cent! In such cases it is very important that you interpret your findings with caution, or better still, let *multivariate* methods help you draw your inferences. Some common multivariate techniques are detailed in Chapter 11.

Finally, as we shall see in Chapter 8, correlations computed between pairs of scores on psychological tests are attenuated, or lowered, as a result of the imperfect levels of reliability inherent in the particular scores that are correlated. Fortunately it is possible to correct obtained correlations for this attenuation.

Summary

In this chapter we discussed the limitations of chi-square as a measure of association between two nominal variables and advocated the use of Goodman–Kruskal's lambda measure of association strength. The Pearson product-moment coefficient was introduced as measure of correlation between two interval-level variables and the procedure for determining confidence intervals around an obtained correlation coefficient was illustrated. A method was also introduced for deciding whether the difference between two correlation coefficients obtained for independent samples is significant. Spearman's rho measure of ordinal correlation was shown to be a derivation of Pearson's r. Finally, problems in interpreting obtained association and correlation were elaborated.

7 Regression Analysis

This chapter is an extension of Chapter 6 in that we will continue to deal with assessing the relationship between pairs of scores. As we saw in that chapter, Pearson's product-moment correlation coefficient allows us to measure the degree of *linear* relationship between two interval-level measured variables. Here we move on to the problem of *predicting* a person's score on one interval-level variable given (*a*) his score on another interval-level variable and (*b*) the previously determined relationship between other individuals' scores on the same two variables.

Regression of *Y* on *X*

What would you predict for subject 10's score on variable *Y*, below?

Subjects	Score on variable *X*	Score on variable *Y*
1	10	100
2	12	125
3	5	30
4	8	90
5	9	85
6	4	20
7	7	60
8	15	140
9	20	170
10	2	?

Intuitively, you may feel that the score will be fairly low, perhaps 10 or 15 or so. Let's plot the relationship between the scores on a scattergraph to see if this will give us any more clues (Figure 7.1).

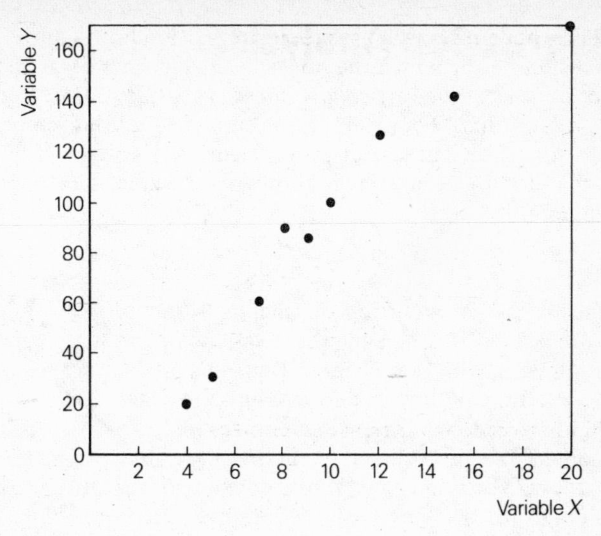

Figure 7.1

Maybe you've now revised your prediction of the Y-score downward somewhat and would now say that it is more likely to be between 0 and 10. Intuitively you may have 'drawn' a straight line through the data to help in your estimation. Indeed a mathematical technique called *regression analysis* does just this.

How does this technique help us to know the *best possible* place to fit a *regression line*? There are, of course, an infinite number of possibilities. Figure 7.2 shows a few.

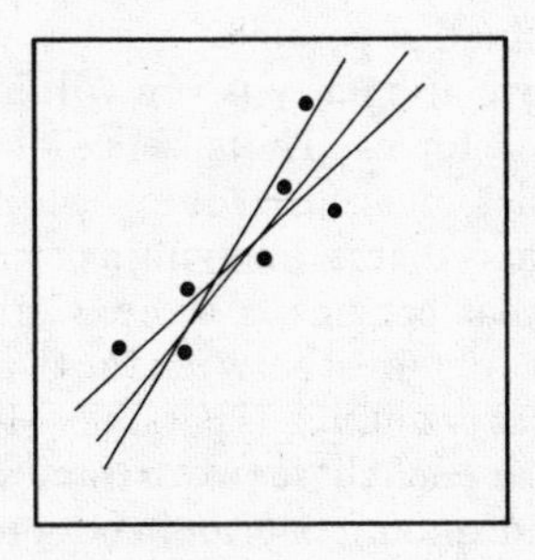

Figure 7.2

The method most often used to fit a regression line is the *method of least squares*, by which the line is located in such a way that *the sum of squares of the vertical distances of the data points from the line is a minimum*. For instance, the sum of squared discrepancies from the line drawn in scattergram 1 (Figure 7.3) is obviously greater than that for the line drawn through the same data points in scattergram 2.

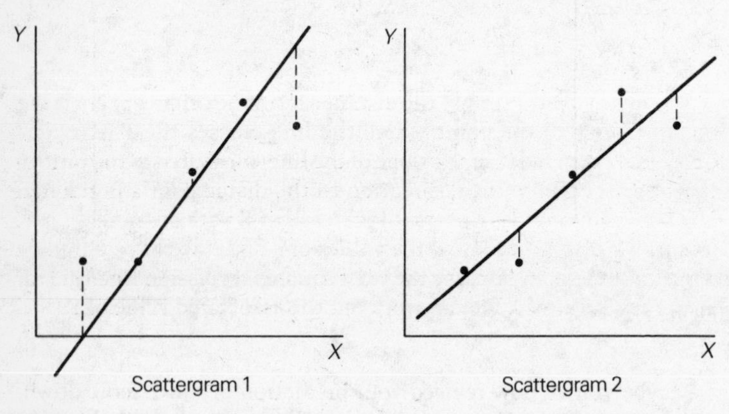

Figure 7.3

In fact, the line drawn on scattergram 2 *is* the least squares regression line. Any change in the position of the line may shorten some of the squared discrepancies *but* it will lengthen others such that the sum of the squared discrepancies is increased.

Basically, the least squares regression line is chosen so that *errors of prediction* are as small as possible.

Prediction and correlation are, as you will have begun to understand, closely related topics. The absence of a linear correlation between two variables means that they bear no systematic linear relation to one another; that is, knowing a person's X-score will not help us to predict that person's Y-score. Conversely, if a perfect linear correlation was obtained between the two variables, i.e. either -1 or $+1$, then perfect prediction is possible. Most obtained correlations, of course, lie between these two extremes and, as we shall see, there will be a certain amount of error in our predictions. It follows, as a rule of thumb, that it is best not to proceed to linear regression if the obtained correlation is *not* significant. It is possible to calculate an index of the amount of error in our predictions, which is called the *standard error of estimate for predicting Y given X*, termed $S_{y/x}$. We will discuss the usefulness of knowing the standard error of estimate later.

Now we turn to the calculations involved in fitting a regression line. We will make use of the general equation for describing any straight line:

$$Y = bX + a$$

The quantity a is simply the distance from the origin or zero point on the Y-axis to the point where the line crosses the Y-axis. The quantity b is an index of the slope of the line, measured as the ratio of the distance in a vertical direction to the distance in a horizontal direction. It describes the rate of increase in Y with an increase in X (Figure 7.4). So, if we know the values of a and b we have uniquely described a line, in our case the least squares regression line, and for any given value of X we can work out the associated value of Y.

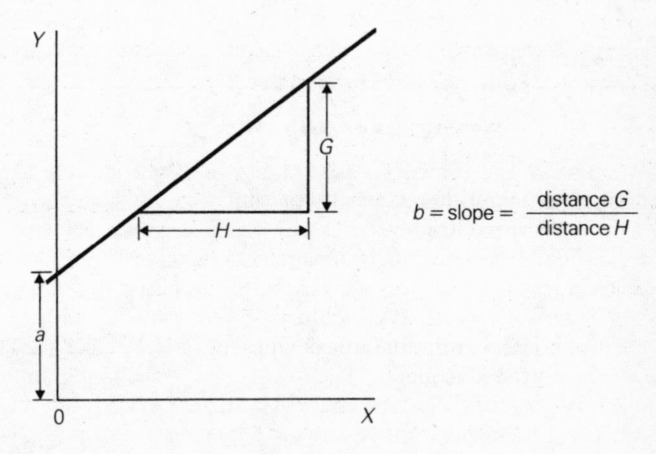

Figure 7.4

Step-by-Step Procedure

Least squares regression equation

	Calculations	**Usual symbols**
Step 1	Lay out the scores for the two variables in pairs (*ignore any pairs with missing values*).	
Step A2	Sum the scores for variable X.	ΣX
Step A3	Divide step A2 by N, the number of pairs of scores.	$\bar{X}$
Step A4	Sum the squares of scores for variable X.	ΣX^2
Step A5	Square the result of step A2.	$(\Sigma X)^2$
Step B2–5	Repeat the above four steps for the Y-scores.	ΣY, $\bar{Y}$, ΣY^2, $(\Sigma Y)^2$

Worked Example

Least squares regression equation

A sample of twelve ex-students were taken at random from City Polytechnic graduates in psychology. For eleven of the students both their final examination percentages and their IQ scores, taken before admission to the psychology course, are known. For the twelfth student only his IQ score is known. The percentage he obtained in the final exam has been mislaid. What 'prediction' would you make of his score?

Step 1

Subject	1	2	3	4	5	6	7	8	9	10	11	12
IQ (variable X)	60	70	80	83	55	90	95	80	65	60	40	70
Final percentage (variable Y)	50	49	63	65	52	68	70	55	45	55	50	?

Step A2 $\Sigma X = 778$. Note, subject 12 has been ignored for now

Step A3 $\bar{X} = \dfrac{778}{11} = 70{\cdot}72$

Step A4 $\Sigma X^2 = 3600 + 4900 + 6400 + 6889 + 3025 + 8100 + 9025 + 6400 + 4225 + 3600 + 1600 = 57764$

Step A5 $(\Sigma X)^2 = (778)^2 = 605284$

Step B2 $\Sigma Y = 622$

Step B3 $\bar{Y} = \dfrac{622}{11} = 56{\cdot}55$

Step B4 $\Sigma Y^2 = 2500 + 2401 + 3969 + 4225 + 2704 + 4624 + 4900 + 3025 + 2025 + 3025 + 2500 = 35898$

Step B5 $(\Sigma Y)^2 = (622)^2 = 386884$

	Calculations	**Usual symbols**
Step 6	Multiply each X-score by its paired Y-score and sum these products.	ΣXY
Step A6	Multiply step A4 by N.	$N\Sigma X^2$
Step A7	Subtract step A5 from step A6 and divide by N, to find the sum of squares of the deviations of X-scores.	$\frac{N\Sigma X^2 - (\Sigma X)^2}{N}$
Step B6–7	Repeat the above two steps for the Y-scores, to find the sum of squares of deviations of the Y-scores.	$\frac{N\Sigma Y^2 - (\Sigma Y)^2}{N}$
Step 8	Multiply step 6 by N.	$N\Sigma XY$
Step 9	Multiply step A2 by step B2.	$\Sigma X \Sigma Y$
Step 10	Subtract step 9 from step 8 and divide this result by N, to find the sum of squares for the cross-products of X and Y.	$\frac{N\Sigma XY - \Sigma X \Sigma Y}{N}$
Step 11	Compute the slope of the regression line by dividing step 10 by step A7.	$b_{y/x}$
Step 12	Compute the intercept with the Y-axis by multiplying step 11 by step A3 and then subtracting its product from step B3 (the regression line passes through $(\bar{X}, \bar{Y})$ and this knowledge allows us to compute the intercept).	$a_{y/x}$

Step 6 $\Sigma XY = 3000 + 3430 + 5040 + 5395 + 2860 + 6120 + 6650 + 4400 + 2925 + 3300 + 2000 = 45120$

Step A6 $N\Sigma X^2 = 11(57764) = 635404$

Step A7 $$\frac{N\Sigma X^2 - (\Sigma X)^2}{N} = \frac{635404 - 605284}{11} = 2738{\cdot}18$$

Step B6 $N\Sigma Y^2 = 11(35898) = 394878$

Step B7 $$\frac{N\Sigma Y^2 - (\Sigma Y)^2}{N} = \frac{394878 - 386884}{11} = 726{\cdot}73$$

Step 8 $N\Sigma XY = 11(45120) = 496320$

Step 9 $\Sigma X\Sigma Y = 778(662) = 515036$

Step 10 $$\frac{N\Sigma XY - \Sigma X\Sigma Y}{N} = \frac{496320 - 483916}{11} = 1127{\cdot}63$$

Step 11 $$b_{y/x} = \frac{1127{\cdot}63}{2738{\cdot}18} = 0{\cdot}412$$

Step 12 $a_{y/x} = 56{\cdot}55 - (0{\cdot}412)70{\cdot}72 = 27{\cdot}41$

	Calculations	Usual symbols
Step 13	To compute the Pearson product-moment correlation coefficient, first multiply step A7 by step B7, find the square root of this product and then divide step 10 by this result; check that this correlation coefficient is significant.	r_{xy}
Step 14	Define the regression equation for your data.	

Regression of *X* on *Y*

In the last section we discussed and illustrated the regression of *Y* on *X*. This regression line was fitted by minimizing the sum of squares of the *vertical* distances of the data points from the regression line, i.e. *those distances parallel to the Y-axis*. However, if we wish to predict with the minimum error values of a variable plotted along the *X*-axis it is best to use a different regression line. This is achieved by fitting a regression line that minimizes the sum of squares of the *horizontal* distances of the data points from the regression line, i.e. those distances *parallel to the X-axis*. This is the regression line of *X* on *Y*. If a perfect correlation of +1 or −1 is obtained between the two variables the regression of *Y* on *X* that of *X* on *Y* will be identical. The lower the correlation between two variables, the more disparate the two regression lines will be.

We will not detail the computational procedure for the regression of *X* on *Y* here, since you will usually be expected to make predictions of scores on *one* variable given scores on another variable *and* the previously determined relationship between a known pairing of scores. To compute the regression equation simply call the variable with the missing scores variable *Y* and call the variable with the known scores variable *X*. Then use the step-by-step procedure!

Step 13 $$r_{xy} = \frac{1127{\cdot}63}{\sqrt{(2738{\cdot}18)(726{\cdot}73)}} = 0{\cdot}799$$

Using Appendix Table II this correlation coefficient is significant.

Step 14 $Y = 27{\cdot}41 + 0{\cdot}412X$

So now we can estimate our missing value of Y, which was paired with an X-value of 70:

$$Y = 27{\cdot}41 + 0{\cdot}412(70)$$
$$= 56{\cdot}25$$

Relationship between Correlation and Regression

Earlier in this chapter we noted that both linear correlation and linear regression have to do with fitting straight lines to data points plotted on a scattergram. *It follows that if the scattergram of your data shows marked curvilinear characteristics (cf. p. 90 of Chapter 6) then application of linear regression will be inappropriate and misleading.* Next we will quantify this qualitative similarity between linear correlation and linear regression.

In fact, a simple relationship can be shown to exist between the correlation coefficient and the slopes of the two regression lines. If the raw scores are translated into standard score form, it can be shown that the best prediction of z_y (i.e. Y expressed in standard score form) is the product of the correlation coefficient (obtained between the other pairs of scores) multiplied by the paired z_x score (i.e. X expressed in standard score form). More formally:

$$z'_y = r\,z_x$$

Similarly, for the regression of X on Y:

$$z'_x = r\,z_y$$

It is also possible to prove that the square root of the product of the two slopes of the regression lines is equal to the correlation coefficient. More formally:

$$r = \sqrt{b_{y/x} b_{x/y}}$$

However, we'll leave the proof to the mathematicians! Next we turn to measurement of error in prediction.

Error in Prediction

Up until now we have treated regression as a *descriptive* statistic, used to describe a complete data set or sample. If we wish to make inferences about the population regression line from the sample data, some assumptions about the data now have to be made. First, we assume that the population distribution of Y is normal no matter what value of X it is paired with. Second, we assume that the standard deviations of these conditional distributions are equal, a condition called *homoscedasticity*.

The first assumption will often be difficult to check since you may not have enough similar X values. It is possible to check for homoscedasticity using the variance-ratio test. In general, unless you have serious doubts to the contrary, proceed directly to computation of the regression equation.

As we mentioned earlier, it is possible to calculate an index of the amount of error in our predictions, called the standard error of estimate. Formally,

$$S_{y/x} = S_y \sqrt{1 - r^2}$$

where S_y is the standard deviation of the Y scores, and r is the correlation between the X and Y scores. The standard error of the estimate varies from 0 to the value of the standard deviation of Y. If our linear prediction is good then the standard error of the estimate will be relatively small.

Intuitively you can see that the larger r^2 is, so that $\sqrt{1 - r^2}$ becomes closer to zero, the better our prediction will be. Indeed, the strength of linear relationship in the data is given by r^2. It is the proportional reduction in variance of Y given the linear rule and a value of X, sometimes called the *coefficient of determination*. For instance, if the value of an obtained correlation coefficient is 0·5 (or for that matter −0·5) then a proportion of 0·25 of the variability in Y

is accounted for by specifying the linear rule and X. If the correlation is 0·6 then 0·36 of the variability is accounted for. Note that the correlation coefficient itself is not a proportion and that a correlation of 0·6 does not indicate a degree of relationship twice as much as a correlation of 0·3.

Let us now consider an example of how we can use knowledge of the standard error of estimate to evaluate predictions. Imagine a normally distributed measure of IQ with a standard deviation, S_y, of 20. Recall that 68 per cent of the area under the normal distribution is included between one standard deviation unit above and below the mean – in other words, 68 per cent of people will possess an IQ somewhere in between 20 points above and 20 points below the mean IQ. Now, if we have a measure of creativity that correlates 0·60 with this measure of intelligence, how much will knowledge of a person's creativity score help us to predict his IQ score?

Since

$$S_{y/x} = S_y \sqrt{1 - r^2}$$

it follows that

$$S_{y/x} = 20 \sqrt{1 - (0{\cdot}6)^2} = 16$$

In terms of prediction, if the IQ/creativity correlation were 0 the standard error of the estimate would, of course, be 20. The obtained correlation of 0·6 has only reduced this to 16. This is not a substantial reduction in the error of prediction and it suggests that predictions should be made with much caution.

However, consider if the correlation between IQ and creativity was higher, 0·95. Then:

$$S_{y/x} = 20 \sqrt{1 - (0{\cdot}95)^2} = 6{\cdot}3,$$

a much more substantial reduction in the error of prediction.

To enable you to calculate the standard error* of prediction use:

$$S_{y/x} = \left(\sqrt{\frac{\text{Step B7 on page 114}}{N - 1}}\right) \left(\sqrt{1 - (\text{Step 13 on page 116})^2}\right)$$

* $N - 1$ is used instead of $N - 2$ as the number of degrees of freedom associated with the sum of squares, to simplify explanation. If we wished to estimate $\sigma^2_{y/x}$ as accurately as possible the sum of squares should be divided by $N - 2$ rather than $N - 1$.

Applying this formula to our example on page 113 gives:

$$S_{y/x} = \left(\sqrt{\frac{726{\cdot}73}{10}}\right)\left(\sqrt{1-(0{\cdot}799)^2}\right) = \left(\sqrt{72{\cdot}67}\right)\left(\sqrt{0{\cdot}362}\right)$$
$$= (8{\cdot}52)(0{\cdot}60) = 5{\cdot}11$$

In this case the standard deviation of Y is 8·52, the first part of our step-by-step formula, and so the reduction in the error of prediction is

$$\frac{8{\cdot}52 - 5{\cdot}11}{8{\cdot}52} = 0{\cdot}40, \text{ or about 40 per cent.}$$

In this case also, predictions should be made with caution.

Multiple Predictors

In real research, rather than in step-by-step examples, you will often find that instead of having a single predictor variable you will have a choice of many. Consider the prediction of final degree marks. Possible predictors could be IQ scores, first-year undergraduate marks, some composite reflecting A-level grades, some measure of achievement motivation, and so forth. An extension of the simple regression of Y on X, called *multiple linear regression*, allows you to analyse the relationship between a single criterion variable and *multiple* predictors. It allows you to evaluate the relative contribution of each predictor variable to the accuracy of prediction. For instance, if the contribution of A-level grades was found to be trivial when used in combination with the other predictors you might then decide to drop it as a predictor. The prediction equation will thus be simplified. In Chapter 11 we describe and give numerical examples of the uses of multiple regression and related multivariate techniques.

Summary

This chapter described the rationale underlying the fitting of a least-squares regression line to pairs of scores from two interval-level variables. Regression of Y on X and that of X on Y were shown to produce different regression lines if a perfect correlation is not obtained between the two variables. Prediction and correlation were shown to be closely related topics. A method of estimating errors in prediction was detailed and, finally, the problem of making a single prediction from multiple indicators was introduced.

8 Questionnaire Design and Analysis

This chapter introduces the topic of questionnaire design and analysis. The main uses of questionnaires – to describe people relative to one another and to make decisions about people – set the scene for a discussion of two crucial issues in questionnaire design, reliability and validity. Computational procedures for estimating reliability are detailed and some simple guidelines for questionnaire wording and response scaling are presented.

Psychological questionnaires are often developed and used as measures of personality, attitudes, intelligence or aptitude. They are a measure of differences between people or, more formally, *individual differences*. Psychological questionnaires are, essentially, standard tasks which give a numerical score of measurement.

Why Use Questionnaires?

1 To describe people

One main use of questionnaires is to *describe* people relative to one another, for example in terms of degree of intelligence or degree of extraversion. These numerical descriptions can then be correlated with other variables to explore patterns of interrelationships. For example, the relationship between degree of intelligence and degree of access to books in the parental home or number of years of post-compulsory schooling could be calculated.

Questionnaire measures can be deliberately devised so that scores, for the population under consideration, are approximately normally distributed such that few people score very high and few score very low on the measure. For instance, on a test of intelligence possible skewness from normality can be prevented by the inclusion of further test items of greater or lesser difficulty. Figure 8.1 shows normally distributed test scores.

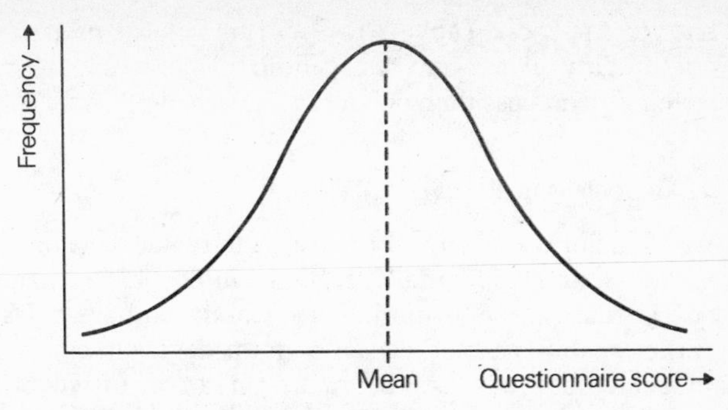

Figure 8.1 Normally distributed test scores

The mean score for the population is called the *norm* and the process by which scores of individuals are compared to the population from which they have been sampled is called *norm-referenced measurement*.

Why bother to devise our questionnaire such that scores are normally distributed? The main reason is because it makes the relative measurement of people very straightforward, since we can make use of z-scores and our knowledge of the normal distribution tables. Recall that:

$$\text{Standard score } z_i = \frac{\text{deviation score x}}{\text{standard deviation}} = \frac{X_i - \mu}{\sigma}$$

where μ = population mean and σ = population standard deviation.

Now, from a questionnaire scoring manual, which accompanies most published questionnaires, we will usually have a good estimate of the parameters μ and σ and so it is easy to calculate the standard score for an observed score.

Let us illustrate this point with an example. Suppose you take an IQ test on which scores for a given population (in this case psychology undergraduates) are normally distributed, and for which μ and σ (or good estimates of them) are known. Suppose that $\mu = 100$ and $\sigma = 15$. You score 130 on the test. What proportion of the population would you expect to have IQs less than yours?

We first express $X_i = 130$ as a z-score:

$$z_i = \frac{130 - 100}{15} = 2$$

Now refer to Appendix Table I. You will find that a proportion of 0·9772 or, roughly, 98 per cent of psychology undergraduates would be expected to score less than or equal to your score!

2 To make decisions about people

The second major use of questionnaires is to make decisions about people. Will that person make a good employee? Is that child retarded? Is that patient suffering from brain damage? Has that student the capability to achieve an honours degree in psychology? Of course all decisions can be made on an intuitive or non-systematic basis, but the argument we will put forward later in this chapter is that the use of standardized psychological measurements will result in better decisions and forecasts.

Figure 8.2 describes a 'typical' relationship with test results and scores on a criterion *to be forecast*. The test results might be, say, scores on a test purported to measure aptitude to understand statistical concepts, whereas the criterion to be predicted might well be overall percentage marks obtained in the final examination of a psychology degree.

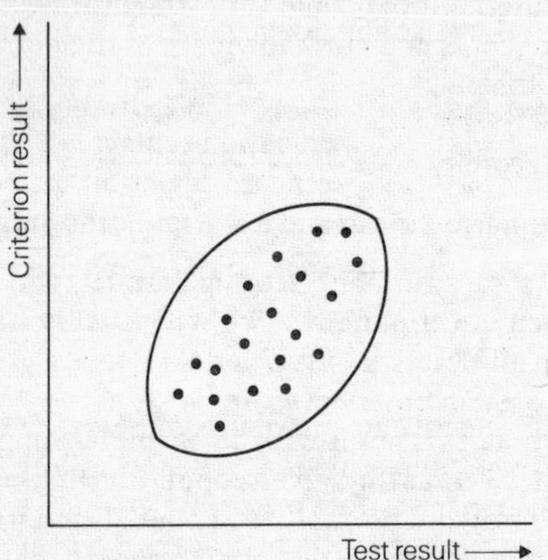

Figure 8.2 'Typical' relationship between test and criterion

We have drawn an oval around individuals' positions on the scattergraph in order to contain all data points. Now, suppose we draw in cut-off scores for the test result and for the criterion result (Figure 8.3).

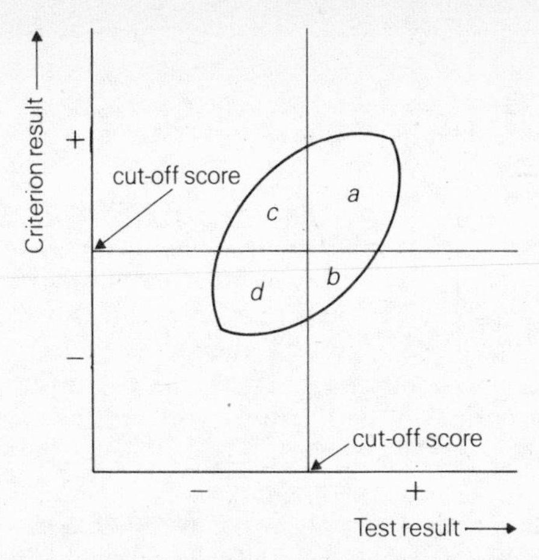

Figure 8.3 Test and criterion relationship with cut-off scores

If a person scores below a cut-off score on the test he 'fails' the test; if he scores above the cut-off score he 'passes'. The same is true of the criterion result. On Figure 8.3, area *a* represents those who passed on the test *and* on the criterion, a *true positive*. Area *b* represents those people who passed on the test *but* failed on the criterion, a *false positive*. Area *c* represents those people who failed on the test *but* passed on the criterion, a *false negative*, whilst area *d* represents those who failed on both the test and the criterion, a *true negative*. Obviously if our test is to be a good predictor of the criterion areas *b* and *c* should be as small as possible. In fact prediction is best when

$$\frac{a + d}{a + b + c + d} = 1$$

Now, if people's results are distributed in a similar manner to those in Figure 8.3 some interesting consequences follow. Imagine you are chief psychologist for a firm recruiting graduates as management trainees. Many more people apply for the vacancies than there are jobs. You use an aptitude test that is a *fair* predictor of future management potential. How should you decide which graduates to employ? The best solution would be to raise the test cut-off score to the level indicated in Figure 8.4.

In this way you *only* accept as trainees people who will be good managers. Area *b*, containing the false positives, has been reduced to

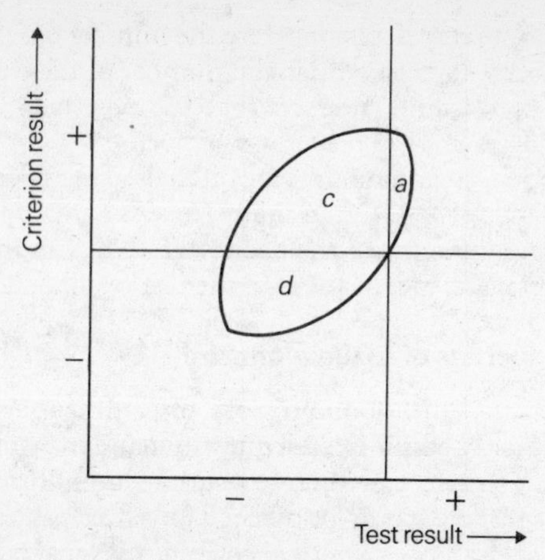

Figure 8.4 Effect of raising the test cut-off score

zero but as a consequence area *c*, containing the false negatives, has increased. In other words, you will engage no unsuitable trainees but you will also reject a large number of potentially suitable trainees. Also note that area *a* has also shrunk, indicating that you will be offering jobs to fewer people.

Figure 8.5 represents the change in a cut-off score that could be

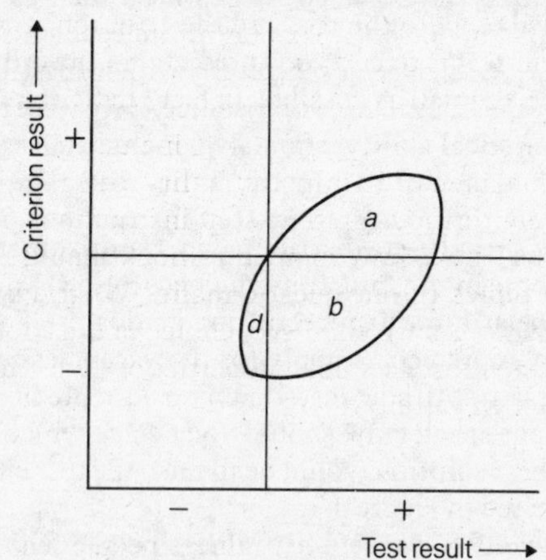

Figure 8.5 Effect of lowering the test cut-off score

applicable to a surgery decision. Here the number of false negatives has been reduced to zero whilst the number of false positives has increased. At the same time *all* the true positives will undergo exploratory surgery.

To summarize, questionnaires are often used to make decisions or predictions about people. Often these decisions are imperfect and so the value of the outcome of the decision also has an impact on the decision to be made, independent of the test result.

Desirable Properties of a Questionnaire

As we have said, questionnaires are often imperfect predictors. Indeed, if a quickly administered questionnaire measure of statistical ability had a perfect predictive relationship to future examination percentage there would be no need for a final psychology exam! Clearly, in this example statistical aptitude may just be one of many abilities necessary for success in psychology examinations. But, anyway, how do we know that the test of statistical aptitude we have chosen to use is the best test of that ability? As you will be beginning to realize, often test and questionnaires are used in psychology to provide an *observable* empirical measurement of an *unobservable concept* that underlies the measured response. A major problem in questionnaire design is *evaluating how well the questionnaire or test represents the underlying theoretical concept.*

Often the theoretical concept is abstract, such as intelligence, creativity, extraversion, introversion or, indeed, statistical ability and there is a problem in deciding exactly which questions should be included in the questionnaire. For instance, if we were to construct a measure of statistical ability, should we include questions on addition, subtraction, mental arithmetic, ability to use statistical tables, ability to follow tortuous step-by-step instructions and ability to 'prove' formulae? If so, how many questions should be devoted to each of these topics in our questionnaire? What, exactly, should each question ask? As we shall see, choice of questions is directly linked to the two most desirable properties of a questionnaire: high reliability and high validity.

Reliability

Reliability concerns the extent to which a questionnaire gives the same score for the same person on repeated trials. It is easy to see that one well-known measuring instrument, the ruler, should give the same measure of a person's height on separate occasions, if the person has not grown or purchased high-heeled shoes in the meantime! Psychological measuring instruments should, ideally, have the property of perfect reliability. However, all measurement has some inherent error such that a measurement of the same quantity will differ from trial to trial. In the case of the ruler, human inaccuracies such as reading heights from positions that are not horizontal with the scale of the ruler will produce error. Unreliability in questionnaires may result from causes such as ambiguous questions that can be interpreted in more than one way, subject boredom and fatigue which prevent the person comprehending the full meaning of a question, or sloppy scoring of the questionnaire by the researcher.

This *random* measurement error sometimes acts to increase a subject's score and sometimes to decrease it; it is totally unsystematic. Obviously, we should attempt to eliminate random measurement errors and so increase the reliability of our questionnaires. At least, we should be able to evaluate the reliability of a particular questionnaire. Next we will discuss ways of estimating reliability. Ways of calculating reliability and ways of increasing the reliability of questionnaires will be dealt with later in this chapter.

1 Re-test method

Simply give the *same* test to the *same* people after a period of time. If the test reliability is high, test/re-test correlations should be high and positive. A person who scores low on the first administration should score low on the second administration. Conversely, a person who scores high the first time should score high the second time. Also, of course, each individual's score should be the same on both occasions.

However, there are problems with this method. First, the method can be expensive and impractical; you will have to keep trace of subjects and have them return to complete the second testing. Second, low correlations may mean that the test is unreliable, but they could also mean that people's attitudes, personality or intelligence have *changed* between testings. Perhaps the process of completing a first test will prompt people to think more deeply about the issues involved. These competing explanations of a low test/re-test correlation are especially difficult to evaluate if the time interval between testings is long. However, if the time interval is short, subjects may remember, and repeat their initial responses. As we saw in Chapter 2, subjects often try to please experimenters!

2 Alternative forms method

This differs from the re-test method in that a *different* but equivalent test is used on the second testing. But, like the re-test method, it does not differentiate true change in a person from unreliability, while the development of equivalent items to be included in the second questionnaire may be troublesome.

3 Split-halves method

Here a longer test is completed on a single occasion. This longer test is divided into two halves before scoring. As with the alternative forms method, however, development of equivalent extra items may require careful thought. Also different ways of dividing the test by first half/second half or odd/even items tend to produce different reliability estimates.

4 Internal consistency method

In this method scores on *every item* of the test are correlated with scores on every other item across a sample of subjects. The average inter-item correlation can then be taken as an index of reliability. Unfortunately, for even a 10-item questionnaire the number of inter-item correlations to be computed and then averaged is 45 and so computer-aided analysis becomes a necessity! Nevertheless, this method would seem to have fewer disadvantages than the alternatives, and it is by far the most popular way used to estimate reliability.

We will detail the computational procedures for estimating reliability later in this chapter. Next we turn to discuss the second most desirable property of a questionnaire: high validity.

Validity

Validity concerns the extent to which a questionnaire measures what it is intended to measure. While reliability focuses on consistency of measurement, validity focuses on the crucial relationship between the (usually) unobservable concept and the measured response. High reliability is a prerequisite for high validity, but high reliability does not by itself imply high validity. Consider the usefulness of a poorly manufactured sub-standard ruler that under-measures by one centimetre in every metre: it will produce highly reliable but totally invalid measurements of distance. The *non-random error* has a *systematic* biasing effect. Validity depends on the extent of non-random error in measurement. Several methods have been devised to estimate the validity of psychological measures.

1 Content or face validity

This measure is purely qualitative and refers to the relationship between the perceived adequacy of the content of the test and the theoretical construct. For instance, a test of mathematical ability designed for primary school children should perhaps include questions on addition, multiplication and division and not exclude questions on subtraction. Similarly, a general test of sociability which places a strong emphasis on the number of friends made within the last year would not have high face validity for first-year undergraduates, since the majority are away from home for an extended period for the first time and will tend to have made more friends than average for their age group. Measurement of face validity has more to do with commonsense and insight than numerical measurement.

2 Criterion-related validity

This is a measure of how well the questionnaire relates to a criterion. The operational measure of criterion-related validity, the correlation between the test and the criterion, is often called the *validity coefficient*. Criterion-related validity might be concerned with predicting a future performance criterion, so-called *predictive validity*, or correlating the questionnaire score with a criterion at the same point in time, so-called *concurrent validity*. Tests used by companies to aid selection of applicants are concerned with predictive validity, whilst the Department of Transport's driving test can be thought of as primarily concerned with concurrent validity, since the instant a learner driver passes the test and removes the L-plates he or she will be treated by other road-users as a competent driver.

Usually, and especially in the social sciences, criteria are disputable. Consider employee selection by aptitude questionnaire. What criterion would be best? The length of time the employee stays with the company? The performance of the employee as judged by superiors? Absenteeism? Speed of promotion within the company? Clearly no single criterion may be appropriate and several criteria may need to be considered. In psychological research, a truly representative criterion may be difficult to pin-point since the concepts underlying the questionnaire are often too abstract. What single or multiple criterion would you consider representative of intelligence or introversion?

3 Construct validity

Construct validity is an attempt to circumvent the problems with establishing criterion-related validity. It is concerned with the extent to which a particular questionnaire relates to other questionnaires in a way consistent with the theoretical predictions which are part of the concept underlying the questionnaire (Figure 8.6).

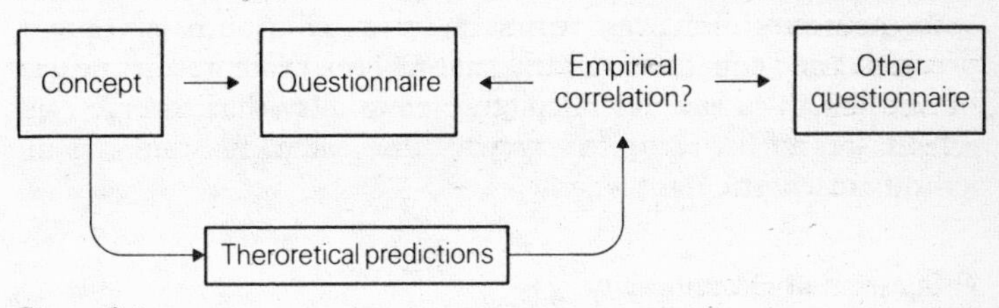

Figure 8.6

For example, the literature on authoritarianism, dogmatism and intolerance of ambiguity assumes that people who are high scorers on scales measuring these concepts see the world 'in black and white' and make extreme judgements and responses. One of the primary characteristics of an individual who is intolerant of ambiguity is need for certainty. Similarly, the authoritarian personality has no time for ambivalence or ambiguity. From these and other theoretical considerations you might expect that scores on questionnaire measures of each of the three will be highly correlated.

Indeed, this has been shown to be the case. Now imagine you have developed a measure of 'decisiveness in decision-making'. Perhaps the theory underlying your measure has aspects in common with the brief outlines of the concepts sketched out above. If it has you should test the construct validity of your measure by correlating your test scores with scores on measures of these concepts to assess *convergent validity*. Alternatively, if you have developed a measure of a theoretical construct that should bear *no* relationship to the above constructs it is possible to evaluate the construct validity of your test by assessing *discriminant validity* – there should be no significant correlations between your measure and measures of the other three constructs.

Clearly, the process of evaluating construct validity involves close theoretical understanding. Sometimes, because of the sparseness of theory surrounding a particular concept, it is impossible to make clear-cut theoretical predictions of correlations. Also, a major problem with the process of construct validation is that the other tests may themselves have low reliability and validity. Indeed, the absence of one or two predicted correlations should not instantly 'write-off' a test you have developed!

Unfortunately, there is no 'cookbook' answer to say which is the best way to assess the validity of a test. Often you will find that when new questionnaires are first reported in the journals, content, predictive and construct validity have all been evaluated in one way or another. It is the *overall* picture of validity that emerges and which should be evaluated by you, the reader, in assessing the validity of a particular test.

Questionnaire Wording

As we shall see in the next section, measures of attitudes and personality often present the respondent with statements and ask for a response which indicates his or her degree of agreement with those statements. For example:

'It is better to come to a decision one way or the other rather than to remain undecided'	strongly agree						strongly disagree

How should you go about constructing such statements? Here are some general rules of thumb:

(*a*) Avoid statements that can be interpreted in more than one way, e.g. 'Decisions have to be taken'. By oneself? By others?
(*b*) Avoid statements that will be answered in the same way by everybody, e.g. 'Sometimes I have to make decisions'; remember you are interested in individual differences!
(*c*) Avoid complicated language, e.g. 'Decision-making is a complex indication of a super-ordinate intellectual capacity'.
(*d*) Keep statements short.
(*e*) Avoid the use of double negatives, e.g. 'Decision-making is not a non-requirement of strong leadership'.

The better constructed the question, the more reliable the questionnaire is likely to be.

Response scales

The simplest method of response scaling was proposed by Rensis Likert in 1932. Simply construct statements and ask subjects to indicate their agreement or disagreement along a five-point scale ranging from, say, 'strongly agree' to 'strongly disagree'. Imagine you have constructed a twenty-question questionnaire on 'aggressiveness' containing such statements as

'Often I feel like hitting people'	strongly agree						strongly disagree

If we count a 'strongly agree' as 5 points and a 'strongly disagree' as 1 point, with three graduations in between, a score of 100 on the questionnaire would indicate the highest level of aggressiveness whereas a score of 20 would indicate the lowest level of aggressiveness – given, of course, that the test was highly reliable and valid. We could also add *reversed* items such as:

'I am a placid person'	strongly agree						strongly disagree

where a 'strongly agree' would score 1 and a 'strongly disagree' would score 5. Such question-reversing would identify 'yea-sayers', those people who tend to acquiesce, or agree, with everything.

Similarly, adding 'filler' statements which will *not* be scored such as:

'I find my work interesting' strongly agree | | | | | | strongly disagree

will tend to confuse people as to the real purpose of the questionnaire, in this case measurement of aggressiveness, and so make it more likely that you will receive an honest response to the 'real' questions. Often questionnaires are constructed to investigate areas which are value-loaded with social desirability. For instance, it is perhaps socially desirable *not* to be seen as being physically aggressive. If respondents realize that your questionnaire is measuring aggressiveness they may 'fake' placidity. It is possible to identify those respondents who are likely to be lying by presenting statements such as:

'I never tell lies'
'I have never stolen anything'

Strong agreement with statements such as these, it is argued, reveals people who are faking their responses to conform to the socially desirable!

Although responses on five-point scales are essentially ordinal-level, in that the psychological distance between each of the points may not be equal, the summated scores for individuals over the entire questionnaire are often normally distributed and are treated as amenable to interval-level statistical analysis.

Similarly, questionnaires using three-point scales:

agree	undecided	disagree

and two-point scales:

yes	no

often lend themselves to interval-level analysis, as we saw in Chapters 3 and 6. If in doubt draw a histogram of your samples' scores and check for strong divergence from normality.

Computation of Reliability Estimates

1 Split-halves method

If we simply correlate scores on one half of the test with scores on the other half the subsequent correlation would be the reliability for each half of the test rather than for the whole test. For this reason a statistical correction must be made in order to estimate the reliability of the total test. The Spearman–Brown prophecy formula is the appropriate statistic:

$$\text{reliability coefficient for the whole test} = \frac{2\,(\text{split-half correlation})}{1 + \text{split-half correlation}}$$

Thus if the split-half correlation is, say, 0·8 then the estimate of reliability for the whole test is:

$$\frac{2(0{\cdot}8)}{1 + 0{\cdot}8} = \frac{1{\cdot}6}{1{\cdot}8} = 0{\cdot}89$$

Notice that the reliability estimate for the whole test is higher than the split-half correlation. The reliability coefficient will vary between 0 and 1, if the split-half correlation is positive. Obviously, it would be pointless to compute the Spearman–Brown prophecy formula if the split-half correlation turned out to be negative!

A more general version of the prophecy formula is:

$$\text{reliability coefficient for the whole test} = \frac{N(C)}{1 + (N - 1)C}$$

where N is the number of times longer the whole test is than the part-test and C is the reliability coefficient for the part-test. This more general formula is useful because it can be shown by rearrangement of the terms in the formula that:

$$N = \frac{\text{desired reliability}\,(1 - C)}{C(1 - \text{desired reliability})}$$

Thus, if a 10-item test has a reliability coefficient of 0·5 then the *estimated* lengthening of the test to result in a desired reliability of 0·8 is:

$$N = \frac{0{\cdot}8(1 - 0{\cdot}5)}{0{\cdot}5(1 - 0{\cdot}8)} = \frac{0{\cdot}8(0{\cdot}5)}{0{\cdot}5(0{\cdot}2)} = 4$$

In words, you would have to add 30 more similar items to your original 10-item test to achieve an estimated reliability of 0·8.

What is an acceptable level of reliability for a test? Carmines and Zeller (1979) argue that reliabilities should not be below 0·8.

2 Internal consistency method

The most popular is *Cronbach's alpha* (α) which states that:

$$\alpha = \frac{n\,(\text{mean inter-item correlation})}{1 + (\text{mean inter-item correlation})(n - 1)}$$

where n is the number of items in the test.

For instance, if the mean inter-item correlation of a 20-item test was found to be 0·63, then the alpha for the 20-item test would be

$$\frac{20\,(0{\cdot}63)}{1 + 0{\cdot}63(20 - 1)} = \frac{12{\cdot}6}{1 + 11{\cdot}97} = \frac{12{\cdot}6}{12{\cdot}97} = 0{\cdot}97$$

Like the Spearman–Brown prophecy formula, Cronbach's alpha varies between 0 and 1 and you should interpret it in the same way, i.e. do not accept a reliability coefficient of less than 0·8. Notice also that if you add extra items to your test and these do not decrease the average of the inter-item correlations it follows that the reliability of your questionnaire will increase. Conversely, if some of the items already contained in your scale have low correlations with the majority of other items it follows that if you exclude these items from your questionnaire then coefficient alpha will increase.

Clearly, calculation of inter-item correlations will enable you to 'purify' your questionnaire, in that items unrelated to the main body of your questionnaire can be eliminated. However, a pattern of low or near-zero correlation may also indicate that there is more than one dimension to your questionnaire. For instance, a scale measuring aggressiveness may contain items related to aggressiveness to one's own family and friends and also aggressiveness to people in authority. Clearly the two sorts of aggressiveness may not be related and so high correlations between the two sorts of items will not be observed. We will deal with this more problematic question of isolating sub-scales later, in Chapter 11 on multivariate analysis.

3 Correction for attenuation

Once you have calculated the reliability of your test using one of the two methods detailed above, you can use this estimate to 'correct' obtained correlations between questionnaires for the unreliability caused by random measurement errors. For instance, if we know the reliability coefficient for each of two questionnaires as well as the correlation between them we can estimate what this correlation would be *if* each questionnaire was perfectly reliable. This procedure is called *correction for attenuation*. Attenuation simply means 'reduction', in this case reduction of the obtained correlation between the two tests due to the less than perfect reliability of each of the tests. The formula to correct correlations for attenuation is:

$$\text{correlation corrected for attenuation} = \frac{\text{obtained correlation}}{\sqrt{(\text{reliability of } Q_1)(\text{reliability of } Q_2)}}$$

For example, imagine that you obtained a correlation between scores on two questionnaires of 0·6 and you knew that the reliabilities of the questionnaires had been estimated at 0·8 and 0·9, respectively. The obtained correlation of 0·6 corrected for attenuation would be:

$$\frac{0{\cdot}6}{\sqrt{(0{\cdot}8)(0{\cdot}9)}} = \frac{0{\cdot}6}{\sqrt{0{\cdot}72}} = \frac{0{\cdot}6}{0{\cdot}849} = 0{\cdot}71$$

In general, the higher the reliability of the questionnaires correlated, the nearer the corrected correlation will be to the obtained correlation. If the reliability estimate for each of the questionnaires is 1·0 then the obtained correlation between them will be equal to the corrected correlation. Because of fluctuations in test reliability estimates it is possible to derive correlations with a value greater than one. The expression

$$\sqrt{(\text{reliability of } Q_1)(\text{reliability of } Q_2)}$$

is an estimate of the maximum possible correlation between scores on the two questionnaires. Correlations that have been corrected for attenuation and which exceed this value indicate such instability.

Attempts to estimate the construct validity of a test by convergent validation with other tests usually result in attenuated correlations, since all tests are to some extent unreliable. For this reason it is extremely important that you read through the manual of any published questionnaire that you intend to use in the convergent validation of a questionnaire that you have developed, and find the reliability estimates. Notice that different reliability estimates are often obtained for different classes of respondent, e.g. males or females, college students or the general public.

Summary

This chapter discussed two major uses of questionnaires, norm-referenced measurement and aiding selection decision-making. Since selection decisions based on questionnaires are often imperfect the values of the outcomes of a decision have an impact on the decision to be made, independent of the test result. Methods of estimating the reliability and validity of questionnaires were evaluated and the split-halves and internal consistency methods were shown to produce good estimates of reliability. Simple rules of thumb were presented for questionnaire wording and response scaling.

9 Parametric Analysis of Variance

In Chapter 4, we described the *simple* experiment, involving a single control group and a single experimental group. In this sort of investigation, we make our causal inferences from the difference between only two sample means. Psychological investigations are rarely this simple. They often involve several groups of subjects, each group being given a different treatment or combination of treatments. These complex designs require the researcher to compare a number of means simultaneously, in order to test for any causal relationship between the independent and dependent variables. Analysis of variance techniques offer us a means of doing this and so are employed in the analysis of data obtained from the more sophisticated experimental designs.

In this chapter we aim to provide you with the necessary knowledge and skills to know when and how to carry out Analysis of Variance (ANOVA). We will start with an explanation of the rationale behind ANOVA and a description of the various models. We will then discuss the advantages and limitations of the ANOVA technique. Finally, we will explain how to compute an ANOVA and the way in which the results should be interpreted.

Although we considered it important to present the computational procedures involved in ANOVA in some detail, we also recognize that these days most of you will use a computer to carry out the actual analysis. Therefore the section on the calculations involved in ANOVA has been presented in a very condensed form. Consequently it may appear complicated and rather off-putting for most of you. However, its aim is as much to help you understand what a computer does with your data during an analysis of variance, as it is to develop competence in manual computation. Hence we go straight to one of the most complicated designs for illustration. Although it is hoped that some of you may be interested to follow in detail the separate calculations involved in the analysis, it is likely that the majority of you will find the essential sections on data presentation and interpretation of results of most use when adopting the technique.

When and Why Do We Use ANOVA?

Analysis of variance techniques should only be applied in situations where the basic assumptions underlying all parametric tests can be either met or overcome (see Chapters 3 and 5). They *can* be applied where the levels of the independent variables form nominal groups (e.g. male and female) or rankable groups (e.g. low, medium and high), as long as the dependent variable has been measured on at least an interval scale. Thus, if your research question concerns the effect on a single (dependent) variable of three or more (independent) types or levels of variables, then you should use an analysis of variance technique.

ANOVA allows you to compare a number of means *simultaneously*. Although *t*-tests are useful in comparing the means from two samples, the application of multiple *t*-tests (of necessity in series) to more complex designs increases the probability of making a Type I error. This is because the number of possible comparisons, and hence separate *t*-tests (e.g. A vs B, B vs C and A vs C), will increase with each additional sample group. Thus, with a one in twenty chance (i.e. $P = 0{\cdot}05$) of making a wrong decision regarding the rejection of the null hypothesis, one of your *t*-tests is quite likely to be in error. Simultaneous group mean comparisons not only reduce the risk of a Type I error, but also utilize all the available data in estimating the population parameters. This should, therefore, result in more reliable estimates.

The main advantages associated with the use of parametric tests have been outlined in Chapters 3 and 5. The increased design scope refers to the number of variables that can be incorporated, the complexity of the design, and consequently the range of questions that can be asked. To take design size first, ANOVA's statistical procedures generalize to any number of independent variables and levels within variables. The more independent variables you have in your design, the more *ways* there are in the analysis. A one-way ANOVA has only one independent variable, a two-way has two and so on. Note that we specifically state *independent variables*, since the designs discussed in this chapter are only concerned with one dependent variable (i.e. they are *univariate*). Although some forms of ANOVA exist which can deal with more than one dependent variable, these are beyond the scope of the present book.

The use of ANOVA rather than multiple *t*-tests also provides scope for an examination of the way in which the effects of two or more variables may *interact*. You may be interested in the possible differential effect of one variable across different *levels* of a second variable. For instance, you may predict that the effects of noise on the performance of a manual skill will vary with the level of difficulty of the particular task presented (e.g. that noise will have its greatest effect on the performance of very difficult tasks). Often, the interactions are of more interest than the separate main effects, in this case 'noise' and 'task difficulty'. Interactions are quite independent of the main effects, so that significant main effects do not imply significant interactions (Figure 9.1). Equally, it is quite possible to find a significant interaction but no significant main effects (Figure 9.2). Notice in Figure 9.1 that when there is no interaction effect the lines on the graphs are parallel. The stronger the interaction the further from parallel they are (Figure 9.3) until the two lines cross over, which indicates the strongest type of interaction (Figure 9.2). Some interpretation is provided with each figure and the use of the interaction tables underneath the individual figures will be explained later on in the chapter.

With respect to design complexity, you should already be familiar with the two basic experimental designs, i.e. *between-subject* and *within-subject* designs. Recall that in between-subject designs there is only one observation or score per subject, so that it is expensive in terms of numbers of subjects that have to be run (and thus time taken to complete). On the other hand, a within-subject design involves each subject giving a response under all conditions, so in many respects the subjects act as their own controls. These designs are consequently less prone to unsystematic error from individual subject variation (i.e. the experiments are less 'noisy'); they are, however, more susceptible to confounding effects like order of treatment or subject fatigue.

ANOVA statistical designs and experimental designs are essentially interdependent. In order for the technique to be able to assess the relative magnitude of the variation from the different sources, the experiment must be designed in a particular way. The various

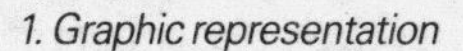

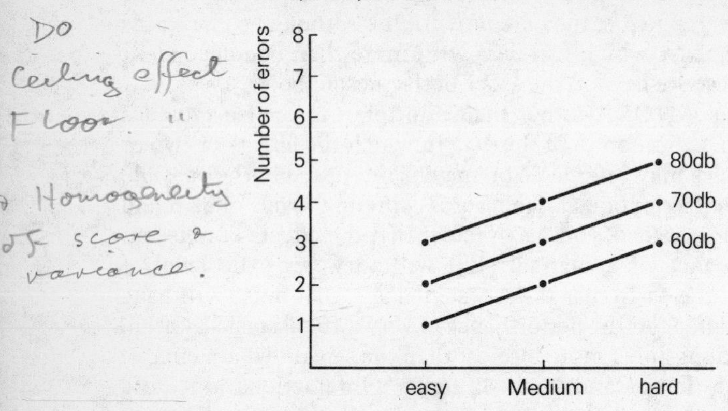

2. Interaction table

TASK DIFFICULTY

		Easy	Medium	Hard	Tot NL
	60db	1	2	3	6
NOISE LEVEL	70db	2	3	4	9
	80db	3	4	5	12
	Tot TD	6	9	12	

Figure 9.1 An example with two significant main effects (noise and task difficulty) but no significant interaction

designs discussed in this chapter will all be examples of different sorts of *factorial* designs, because these are the types most commonly used in experimental psychology. The variables included in the design are called *factors* and, as already noted, we can talk of different *levels* within a factor. In a factorial design the factors are *hierarchically* organized. This means that all levels of each factor are crossed, or combined, with all levels of the other factors. This can be represented by a factorial matrix (see Tables 9.3, 9.4 and 9.5 on pp. 157–9).

Factorial designs can involve between-subject variables (simple factorial), within-subject variables (repeated measures) or both types

1. Graphic representation

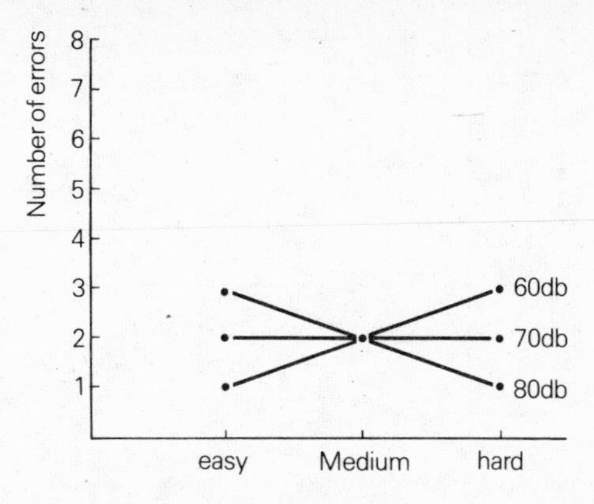

2. Interaction table

		TASK DIFFICULTY			
		Easy	Medium	Hard	Tot_{NL}
NOISE LEVEL	60db	1	2	3	6
	70db	2	2	2	6
	80db	3	2	1	6
	Tot_{TD}	6	6	6	

Figure 9.2 An example with two non-significant main effects (noise and task difficulty) and a significant interaction

of variables (split-plot). Your choice of design will depend on several criteria. The kind of information you are interested in, via your choice of variables, could constrain the nature of your design. For example, if you wish to examine changes over time on a variable, then a within-subject design would be the most appropriate. Equally if you wish to examine a particularly sensitive effect (e.g. diurnal rhythms on memory), the effects may be masked out by 'noise' if you adopt a between-subject design. Of course there are some variables (particularly individual-difference or organismic variables) which demand a between-subject design.

In addition to your choice of variables, the inferences you may

1. Graphic representation

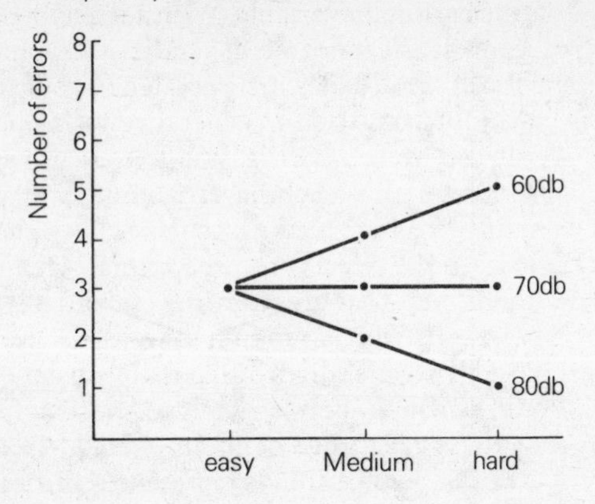

2. Interaction table

		TASK DIFFICULTY			
		Easy	Medium	Hard	Tot *NL*
NOISE LEVEL	60db	3	4	5	12
	70db	3	3	3	9
	80db	3	2	1	6
	Tot *TD*	9	9	9	

Figure 9.3 An example with one significant main effect only, and a significant interaction

wish to draw from your data can also restrict your design choice. This is because ANOVA can be applied to two different sampling situations, each requiring the employment of a slightly different statistical model. The model you adopt will depend on the way in which your treatments have been selected, and the sort of inferences you wish to make from your data. There are basically two different ANOVA models, with a third model representing a combination of the other two. The three models are referred to as the *fixed-effect model*, the *random-effect model* and the *mixed model*.

In a fixed-effect model, all the variables or levels of variables, of interest are actually represented in the design. Sex would naturally be a *fixed effect*, because male and female exhaust all the possible categories of this particular variable. With a fixed model, inferences drawn from your data cannot be generalized beyond the specific variables and levels of variables incorporated in your study. In some situations you would not wish to generalize your findings further. For example, if you were investigating the effectiveness of three different schemes for the teaching of reading, you would not be interested in the effect of reading schemes in general, only the differences between the effects of the particular schemes chosen.

However, there may be a situation where variables included in your experiment constitute a random sample of a much larger set. You may then wish to generalize to all possible instances or levels of a variable, rather than to be restricted to those actually studied. Variables like these are called *random effects*. An example of a random effect might be the schools represented in an investigation. If these schools were selected at random then you could make inferences from your sample to the wider population of similar schools. With respect to choice of design, fixed-effect and random-effect statistical models can be used in both between-subject and within-subject designs.

Because ANOVA is applied to factors represented by at least two values or levels, a situation where *all* the effects can be considered random is rather rare. In most cases, one or more factors will have fixed levels, the remaining factors being sampled. Each individual observation is then seen as a sum of both fixed and random effects i.e. *mixed effects*. The mixed statistical model adopted in these situations is thus a combination of the other two. An example of a situation requiring a mixed model could be an investigation into possible differences between males and females (fixed) in their enjoyment of 'cowboy' films (random).

Mixed models are commonly used in within-subject designs where the subjects themselves are viewed as a variable (albeit a nuisance variable rather than one you set out to examine). Since the subjects are usually randomly sampled they are a special case of a random effect. Therefore, in situations where the variables of interest may be fixed but the design is within-subject, you would need to adopt a mixed model. The computational examples in this chapter will represent either fixed-effect models, or mixed models where the random effect is the subject.

Certain practical considerations are also likely to influence your design choice. Not least of these are the number of available subjects and the amount of time at your disposal.

The flexibility of the analysis of variance techniques has meant that various computational procedures have also been devised to deal with data from other complex experimental designs (e.g. latin squares). These statistical techniques have also been developed to ask further, more specific, questions of the data (e.g. trend tests and planned comparisons).

What is ANOVA?

The rationale behind analysis of variance will be presented and illustrated through a discussion of a one-way ANOVA. We will then demonstrate how this rationale is translated into the computational formulae, and extend this to a two-way example. Much of this discussion relates back to the statistical concepts introduced in Chapter 5. The keen student in search of more statistical detail should consult Hays (1973) or Winer (1971). Other, more simplified, accounts can also be found in Shavelson (1981), Howell (1982) and Plutchik (1983).

From the title 'Analysis of Variance' it is hardly surprising that the concept of variability is central to this method of determining and quantifying the various effects produced in an investigation. Let us consider how an analysis of variance between groups can reflect differences between means. Note that the *variance* of a group of scores represents a measure of their variability, and is calculated by the formula

$$S^2x = \frac{\sum (X - \bar{X})^2}{N - 1}$$

Where X represents each individual score, $\bar{X}$ the sample or group mean and N the number of observations or scores in the sample.

Assume that you have three groups of scores such as those in Table 9.1. These are hypothetical scores obtained under three levels of a particular treatment ($A1$, $A2$ and $A3$). As you can see, two of the groups of scores are substantially the same, in that their means and variances are very similar.

If you combined the two similar groups you would expect the variance of the combined group to be equal to the individual variances of each of the two groups when separate. You haven't,

Table 9.1. Raw data, means and standard deviations of three hypothetical samples

	Sample $A1$	Sample $A2$	Sample $A3$
	6	7	1
	8	9	3
	10	11	5
	12	13	7
	14	15	9
ΣX	50	55	25
$\bar{X}$	10	11	5
S^2	10·00	10·00	10·00

after all, altered the variability of the scores by combining them. Likewise, the mean of the combined group should be very similar to the separate means. Note that if you combine $A1$ with $A2$ and calculate the new mean and variance then $\bar{X} = 10{\cdot}5$ and $S^2 = 9{\cdot}17$.

However, since the third group has a much lower mean, despite having the same variance as the other two, the addition of this group of scores to those of the other two would result in an increase in variability. In other words, the range and thus spread of scores will be greatly increased by adding the third group of scores (from 6 to 15 to 1 to 15). Again from the table, if you combine $A1$, $A2$ and $A3$ and calculate the new mean and variance then $\bar{X} = 8{\cdot}67$ and $S^2 = 15{\cdot}95$.

It is important to note that in our particular example the separate groups had equal variances. An increase in variability could thus only result from a difference between the group means. Hence the importance of the homogeneity of variance assumption in the use of ANOVA. If this assumption is not met, it is difficult to decide whether an increase in variability is due to a difference between the group means or simply a reflection of the differences in variance. However, as we saw in Chapter 5, parametric tests like ANOVA tend to be 'robust' enough to accept some differences in sample variance, as long as the samples are of equal size.

The logic behind ANOVA is derived from statistical theory relating to *populations* rather than *samples* but, as we have already seen, under certain conditions sample statistics can provide unbiased estimates of population parameters (the necessary conditions for making this assumption have already been discussed in Chapter 5). Therefore the theory can be applied to those samples that have been correctly drawn from their populations. However, we know that when dealing with samples, sampling variation alone is likely to produce differences between the sample means. Therefore some variation will occur, regardless of whether the null hypothesis or the alternative hypothesis is correct.

If we return to our example, the lower mean of the third group of scores could have been due to our treatment effect, but it could equally have resulted from sampling variation. In other words, we have two possible sources of variation. When hypothesis-testing, therefore, you need to determine whether the overall effect obtained is due to sampling variation (error) alone, or to an effect of your treatment *plus* sampling error. In order to do this you need to calculate the relative magnitudes of the effects resulting from sampling error and the effects resulting from your treatment. Unfortunately, it is not possible to separate out the error from the true treatment effect directly, because the overall treatment effect obtained will equal a true treatment effect plus error. However, if the magnitude of the error could be determined by an independent method, then the overall treatment effect could be divided by this *error term*. The ratio obtained would then reflect the size of the true treatment effect, since the bigger the ratio, the greater is the contribution of the particular treatment to the results. The way in which we calculate this ratio is outlined below.

Since sampling error is expressed in terms of the variance of the population, if we assume homogeneity of variance (i.e. $\sigma^2{}_1 = \sigma^2{}_2 = \sigma^2{}_3$), this is the same as saying that the variance of any one of our samples would be a good estimate of the variance of the population (σ^2). To get a better estimate, instead of using the variance of only one sample, we could use the mean variance of the combined sample. The sampling error (or variance) could then be expressed as

$$\sigma^2 e = \frac{S_1{}^2 + S_2{}^2 + S_3{}^2}{k}$$

where k = total number of samples (treatments) and S^2 = variance of each sample.

Note that this is the case *whether the null hypothesis is true or false*, because of the homogeneity of variance assumption. If, however, the null hypothesis is true (that is our three samples were in fact drawn from the same population), the central limit theorem will apply. This states that the variance of means drawn from the same population equals the variance of the population divided by N.

So, if all our samples are representative of the population from which they were all drawn, the central limit theorem can be used to provide us with a second estimate of the sampling error (or variance) ($\sigma^2 e$):

$$\text{If } \frac{\sigma^2 e}{n} = S^2\bar{X}, \text{ then } \sigma^2 e = nS^2\bar{X}$$

where n = sample size and $S^2\bar{X}$ = the variance about the mean.

Consequently, we are able to calculate two estimates of our sampling error (or variance). One of these is independent of our hypothesis, and the other is only a good estimate under the null hypothesis. We call the first estimate the error variance, or mean square error (ERROR.MS) and our second estimate the treatment variance, or mean square treatment (TREAT.MS). Remember that the treatment variance estimate will include the true treatment effect plus the error. Therefore, the greater our treatment effect, the greater the disparity will be between our two estimates. Alternatively, when the null hypothesis is true our two estimates should be about equal. The F-value or ratio is calculated by dividing the mean square of the treatment effect by the mean square of the error effect. Obviously if there is *no* treatment effect (i.e. the null hypothesis is true) the resulting *F*-value or ratio would equal 1: the greater the value, the greater the treatment contribution. In theory, it is not possible to have an *F*-value of less than 1, but as the calculated effects are only estimates, values of less than 1 do sometimes occur. To determine whether your particular *F*-value is significant, you need to assess its probability of occurrence in the *F*-distribution. Like the *t*-distribution, the *F*-distribution is a family of distributions. The particular distribution that you need to refer to depends on the number of degrees of freedom associated with your treatment and error effects (see pp. 175–8 for further details). Tables of *F*-values can be found in the Appendix.

The analysis of variance outlined above is based on two fundamental assumptions. First, that the two estimates of sampling error or variance are independent, as a result of *independent* sampling and treatment effects. This independence should be achieved through random sampling and random assignment. Second, it is assumed that the overall treatment effect represents a *sum* of the error plus treatment effect rather than, for example, a product. This is referred to as the *additivity assumption*, and in ANOVA it is made in addition to the general assumptions underlying all parametric tests.

Let us summarize the reasoning behind ANOVA in less statistical language. ANOVA compares the amount of variance within a treatment group with the amount of variance between treatment groups. The *within-group* variance is attributed to sampling variation only, since all members of a particular group are treated alike. The variability *between groups* is attributed to both a treatment effect and a sampling error. If the variance between groups is considerably greater than the variance within groups, then this difference can be attributed to the treatment effect. One obvious consequence of the above, is that if you can reduce the within-group error you are less likely to mask out any treatment effect through excessive subject variability or 'noise'. A well-controlled experiment will attempt to reduce this unsystematic error (see Chapter 4).

Having explored the rationale in general terms we will now attempt to translate this into computational formulae. ANOVA is essentially concerned with the *partitioning of variance*. In our example, ANOVA partitions the total amount of variance present in the data into the treatment variance and error variance. These were our two main *sources* of variance. Thus, for our particular example the partitioning of variance can be represented diagrammatically in Figure 9.4.

The ANOVA computations involved are initially concerned with calculating *sums of squares*. A sum of squares is the sum of the squared deviations about the mean ($\Sigma(X - \bar{X})^2$), or some multiple of this. If we recall that the formula for the total variance in a sample is

$$S^2x = \frac{\Sigma(X - \bar{X})^2}{N - 1}$$

we can see that the numerator represents the sum of the squared deviations (i.e. the sum of squares). Sample variance can also be calculated by the formula:

$$S^2x = \frac{\Sigma X^2 - (\Sigma X)^2/N}{N - 1}$$

In this second expression the numerator ($\Sigma X^2 - (\Sigma X)^2/N$) again

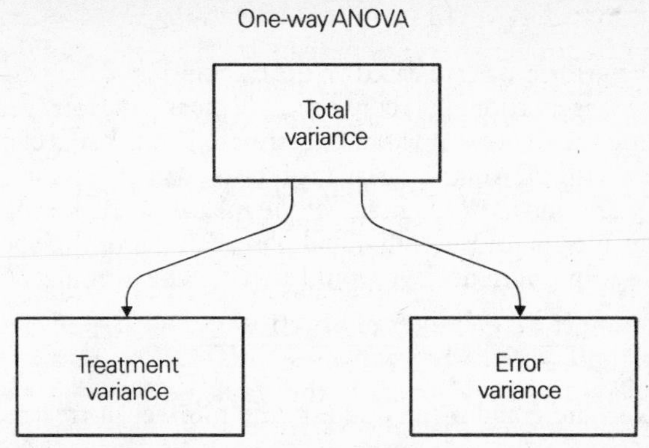

Figure 9.4 The partitioning of variance; the total sum of squares and degrees of freedom can be partitioned in the same way

represents the sum of squares. In both formulae the denominator $(N - 1)$ equals the number of *degrees of freedom* (DF), where N is the total number of scores in the sample.

Therefore, if we return to our example (Table 9.1) and combine all three samples (A1, A2 and A3) into one large sample, we could then calculate the total amount of variance obtained by using the second formula above. However, as we are simply interested in the relative magnitude of the treatment variance and the error variance, we can calculate the numerator (i.e. the total sum of squares) and the denominator (i.e. the total degrees of freedom) separately. Thus

$$\text{Total sum of squares (TOT.SS)} = \Sigma X^2 - \frac{(\Sigma X)^2}{N} \text{ with } N - 1 \text{ DF}$$

where X equals the individual observations or scores and N the total number of scores. The second part of the formula $((\Sigma X)^2/N)$ is referred to as the *correction term* (CT) and is used in most of the ANOVA calculations.

For our example:

$$\begin{aligned} \text{TOT.SS} &= 6^2 + 8^2 + 10^2 + \ldots + 9^2 - \frac{(6 + 8 + 10 + \ldots + 9)^2}{15} \\ &= 1350 - \frac{130^2}{15} = 1350 - 1126{\cdot}67 \\ &= 223{\cdot}33 \end{aligned}$$

and TOT.DF $= 15 - 1 = 14$.

The above formulae are based on the calculation of the variance of individual observations (X) about a sample mean ($\bar{X}$). However, when computing the treatment variance estimate, ANOVA is concerned with the calculation of the variance of the treatment or group means about a *grand mean* or the *mean of the means* ($\bar{G}$). If we substitute treatment means for the individual observations in the above formula for sample variance, we would arrive at the formula:

$$S^2x = \frac{\Sigma(\bar{X} - \bar{G})^2}{k - 1} \text{ or } \frac{\Sigma(\Sigma X)^2/n - (\Sigma X)^2/N}{k - 1}$$

where $\bar{G}$ = the grand mean and k = the number of treatments or groups.

Despite ANOVA being essentially concerned with differences between means, the actual computational procedures involve the use of totals rather than group means. This is because totals are more convenient to use and it makes no substantial difference to the computation since totals and means are linearly related (i.e. $\bar{X} = \Sigma X/n$ or T/n). Thus, with group totals substituted for the means in the second formula, the formula for the *treatment variance estimate (mean square treatment)* becomes

$$\text{TREAT.MS} = \frac{\Sigma T^2/n - (\Sigma X)^2/N}{k - 1}$$

where T = group totals and n = number of observations in each group.

If we calculate the numerator (TREAT.SS) of the equation first, from our example:

$$\text{TREAT.SS} = \frac{T_{A1}{}^2 + T_{A2}{}^2 + T_{A3}{}^2}{n} - \frac{(\Sigma X)^2}{N} \text{ with } k - 1 \text{ degrees of freedom}$$

$$= \frac{50^2 + 55^2 + 25^2}{5} - \frac{130^2}{15}$$

$$= \frac{6150}{5} - 1126{\cdot}67$$

$$= 103{\cdot}33 \text{ with } 3 - 1 = 2 \text{ DF}.$$

To obtain the mean square treatment, we now divide the TREAT.SS by its associated degrees of freedom ($k - 1$), where k represents the number of treatments or groups. In our example:

$$\text{TREAT.MS} = \frac{\text{TREAT.SS}}{\text{TREAT.DF}}$$

$$= \frac{\text{TREAT.SS}}{k - 1} = \frac{103 \cdot 33}{2} = 51 \cdot 66$$

The next step in our ANOVA computation is to calculate the estimate of error variance (mean square error). As we can see from Figure 9.1, the total variance is equal to the sum of the treatment and error variances. Therefore, if we have already calculated both the total and the treatment sums of squares then the error sum of squares can be calculated by subtracting the treatment sum of squares from the total sum of squares (i.e. ERROR.SS = TOTAL.SS − TREAT.SS). This is referred to as calculation by *residual*.

Thus, for our example:

$$\text{ERROR.SS} = 223 \cdot 33 - 103 \cdot 33 = 120$$

The error sum of squares can then be converted into the mean square error by dividing the ERROR.SS by its associated degrees of freedom. The degrees of freedom are also calculated by residual (i.e. ERROR.DF = TOT.DF − TREAT.DF). From our earlier calculations,

$$\text{ERROR.DF} = 14 - 2 = 12$$

Therefore

$$\text{ERROR.MS} = \frac{\text{ERROR.SS}}{\text{ERROR.DF}}$$

$$= \frac{120}{12} = 10$$

In practice, we tend to calculate all the sums of squares first along with their degrees of freedom. We subsequently obtain the mean squares by dividing the former by the latter. Finally, we calculate the *F*-ratio. From our earlier discussions, we noted that the *F*-ratio is determined by dividing the TREAT.MS* by the ERROR.MS so that

$$\text{F-R} = \frac{51 \cdot 66}{10} = 5 \cdot 17$$

In order to assess whether this *F*-value is significant, you would then refer to the *F*-distribution table.

* We have used the terms mean square and variance estimate synonymously. In later sections we adopt the term mean square only.

So far, we have discussed a simple (one-way) ANOVA design having only one variable or treatment. Later on in the chapter we will be discussing more complex designs, some of which have more than one independent variable or treatment. In cases where there are two independent variables, you will obviously need to calculate two treatment sums of squares (one for each variable). In addition there will also be an *interaction effect* to calculate.

How do we compute the sums of squares and mean squares of these interactions? Some hypothetical data obtained by using a two-way design is set out in Table 9.2. Notice the additional variable or treatment (treatment B).

We start as usual by computing the total sum of squares:

$$\text{TOT.SS} = 2^2 + 1^2 + 2^2 + \ldots + 3^2 - \frac{(102)^2}{30}$$
$$= 440 - 346{\cdot}80 = 93{\cdot}20 \text{ with } 30 - 1 = 29 \text{ DF}$$

followed by the two treatment sums of squares:

$$\text{TREAT(A).SS} = \frac{32^2 + 31^2 + 39^2}{10} - 346{\cdot}80$$
$$= 3{\cdot}80, \text{ with } 3 - 1 = 2 \text{ DF}$$

$$\text{TREAT(B).SS} = \frac{57^2 + 45^2}{15} - 346{\cdot}8$$
$$= 4{\cdot}8, \text{ with } 2 - 1 = 1 \text{ DF}$$

Table 9.2. Hypothetical data table for a two-way ANOVA

Treatment 1	*A*1		*A*2		*A*3	
Treatment 2	*B*1	*B*2	*B*1	*B*2	*B*1	*B*2
Data	2	4	8	2	3	3
	1	2	7	1	4	3
	2	3	2	2	5	5
	3	5	3	1	6	4
	4	6	4	1	3	3
Total (*AB*)	12	20	24	7	21	18
			Grand total (*G*) or		$\Sigma X = 102$	
					$N = 30$	

The calculation of the interaction sum of squares (INT.SS) is not very different from that of treatment sums of squares. You still need to square totals, but the totals used are those of each variable at each level of the other variable (i.e. $T_{A1B1} = 12$, $T_{A1B2} = 20$, $T_{A1B3} = 24$, $T_{A2B1} = 7$, $T_{A2B2} = 21$ and $T_{A2B3} = 18$). These totals should then be squared and summed:

$$= \frac{12^2 + 20^2 + 24^2 + 7^2 + 21^2 + 18^2}{5} - 346{\cdot}80$$

$$= 40$$

However, each of these totals is also, in part, reflecting the separate treatment effects. Therefore, these main effects should be subtracted to leave us with a pure interaction sum of squares:

$$\text{INT.SS} \quad = 40 - (3{\cdot}8 + 4{\cdot}8) = 31{\cdot}4$$

The interaction effect should not be treated any differently from a main effect, so we need to know how many degrees of freedom are associated with our INT.SS. The INT.DF = TREAT(A).DF × TREAT(B).DF. In our example,

$$\text{INT.DF} \quad = 2 \times 1 = 2 \text{ degrees of freedom.}$$

We still need to calculate our error sum of squares. This is done in a similar way to our first example but in addition to the two main effects sums of squares, the interaction sum of squares has also to be subtracted from the total sum of squares. Thus

$$\text{ERROR.SS} \quad = 93{\cdot}2 - (3{\cdot}8 + 4{\cdot}8 + 31{\cdot}4) = 53{\cdot}2$$

and

$$\text{ERROR.DF} \quad = 29 - (2 + 1 + 2) = 24$$

The mean square for the interaction is calculated in the normal manner. Therefore, in our example the four mean squares would be

$$\text{TREAT(A).MS} = \frac{3{\cdot}8}{2} = 1{\cdot}9$$

$$\text{TREAT(B).MS} = \frac{4{\cdot}8}{1} = 4{\cdot}8$$

$$\text{INT.MS} = \frac{31{\cdot}4}{2} = 15{\cdot}7$$

$$\text{ERROR.MS} = \frac{53{\cdot}2}{24} = 2{\cdot}22$$

We can now compute the *F*-ratio as normal for the two treatment conditions and their interaction:

$$\text{F-R TREAT(A)} = \frac{1{\cdot}9}{2{\cdot}22} = 0{\cdot}86$$

$$\text{F-R TREAT(B)} = \frac{4{\cdot}8}{2{\cdot}22} = 2{\cdot}16$$

$$\text{F-R INT(A} \times \text{B)} = \frac{15{\cdot}7}{2{\cdot}22} = 7{\cdot}07$$

Later on in the chapter we will explore these formulae more fully, and explain how to check the significance of your *F*-ratios.

How to Apply ANOVA to the More Complex Experimental Design

In this part of the chapter, we will describe the important procedures and calculations involved in using ANOVA. We will restrict the discussion to the fixed-effect and mixed-effect statistical models as used in the simple factorial, repeated measures and split-plot designs with *equal* cell sizes.

In the last section we demonstrated how the ANOVA rationale can be translated into computational formulae. From the two examples described (i.e. a one-way and two-way) it can be seen that many of the calculations are common to both designs. For the total sum of squares, the main effects and the interaction effects the computational procedures are essentially the same, regardless of the size or type of ANOVA design. Consequently, our approach to the description of how to use ANOVA involves the adoption of a modular system, which is seen as a more economical approach. Each module will therefore represent a specific *set* of actions or calculations, rather than a single step or formula. Some modules (I, II, V and VI) are common to all basic designs, whereas others (III and IV) are only required by some of them. These latter modules are concerned with the computational procedures involved in between-subject designs (Module III) and within-subject designs (Module IV). The split-plot design contains both within-subject and between-subject components, so incorporates both Module III and Module IV.

Although the split-plot is the most complex design, it is the example we have chosen by which to introduce the various computational modules, because it incorporates all of them. The two modules which do not involve any calculations will be discussed in more general terms. In addition, modular and worked examples of the other two designs (simple factorial and repeated measures) are provided at the end of the section.

Module I: *Data Presentation*

The first step in the analysis of your data is to tabulate them in the form of an appropriate *data matrix*. A matrix is a hierarchical organization of the data, which reflects the different levels within factors and across factors. Table 9.3 is an example of the way in which data from a 2 × 3 simple factorial (two ways and between-subject) may be presented. The study could be investigating the effects of sleep deprivation and article length on the number of errors made by copy typists.

Table 9.3. Factorial matrix for our example of a 2 between-subject (simple factorial) design

Article length (B)	Sleep-deprived ($A1$)			Non-sleep-deprived ($A1$)		
	$B1$	$B2$	$B3$	$B1$	$B2$	$B3$
	1	4	9	1	2	5
	3	4	8	0	4	4
	2	5	8	1	3	4
	1	5	7	2	4	5
	1	4	7	2	3	5
Total (AB)	8	22	39	6	16	23

G or $\Sigma X = 114$
$N = 30$

In the above example, the groups of scores under each different *condition* (particular combination of sleep treatment and article length treatment) constitute the *cells* of the design. We therefore have six cells (2 × 3) and each cell contains the scores of five different subjects. Since it is a between-subject design, each subject provides one score only (X). Hence the total number of observations (N) = 2 × 3 × 5 = 30. The total number of typing errors can now easily be computed for each of the six conditions (columns) and summed together to provide a grand total (G). For a within-subject (repeated measures) design, the data matrix required would be slightly different (see Table 9.4). This example could represent the sort of data obtained from an investigation into the effects of noise and practice on the number of errors made by copy typists.

Table 9.4. The factorial matrix for our example of a 2 within-subject (repeated measures) design

	60 dbs (C1)					85 dbs (C2)					
Trials (D)	D1	D2	D3	D4	(Tot.)	D1	D2	D3	D4	(Tot.)	S.Tot.
S1	2	4	5	6	(17)	1	4	6	9	(20)	37
S2	4	5	6	7	(22)	2	4	7	10	(23)	45
S3	3	3	6	8	(20)	3	3	7	8	(21)	41
S4	2	4	5	8	(19)	3	5	8	10	(26)	45
S5	2	3	7	8	(20)	2	5	7	10	(24)	44
Total	13	19	29	37	(98)	11	21	35	47	(114)	212

In this sort of design, 'subjects' are treated as one of the factors under investigation. Each subject represents *one* level within this factor, and all the levels (subjects) are presented vertically down one side of the matrix. This is because each subject undergoes, and so provides a score, under *all* conditions (in our example trials and noise). This organization allows the various scores from each subject to be presented in rows across the page. Again, the groups of scores under each condition constitute the eight (2 × 4) cells of the design, and the total number of observations (N) = 2 × 4 × 5 = 40. Two different groups of totals need to be calculated in this sort of design. These are the column totals (total number of typing errors made under each condition) and the row totals (total number of errors made by each typist). The sum of both of these groups of totals should equal the grand total (G).

The matrix for a split-plot design is a combination of the between-subject and within-subject matrices. An example of this type of design matrix is illustrated in Table 9.5. The scores could again

Table 9.5. Factorial matrix for our 2 between-subject and 2 within-subject split-plot example

		Within-subject variables						
Noise		C1 (60dbs)			C2 (85dbs)			
Trials		D1	D2	(Tot.)	D1	D2	(Tot.)	S.Tot.
Between-subject variables								
A1 (non-sleep-deprived)								
B1 (short article)	S1	2	2	(4)	8	8	(16)	20
	S2	1	1	(2)	9	7	(16)	18
	S3	3	2	(5)	7	6	(13)	18
	S4	2	1	(3)	8	7	(15)	18
	S5	3	2	(5)	7	5	(12)	17
	Total	11	8	(19)	39	33	(72)	91
B2 (long article)	S6	3	2	(5)	4	4	(8)	13
	S7	4	3	(7)	3	3	(6)	13
	S8	3	3	(6)	2	2	(4)	10
	S9	2	2	(4)	4	2	(6)	10
	S10	4	3	(7)	5	3	(8)	15
	Total	16	13	(29)	18	14	(32)	61
A2 (sleep-deprived)								
B1 (short article)	S11	6	5	(11)	9	8	(17)	28
	S12	5	5	(10)	8	7	(15)	25
	S13	6	4	(10)	6	6	(12)	22
	S14	4	2	(6)	5	5	(10)	16
	S15	5	3	(8)	8	7	(15)	23
	Total	26	19	(45)	36	33	(69)	114
B2 (long article)	S16	4	3	(7)	9	9	(18)	25
	S17	3	2	(5)	10	8	(18)	23
	S18	4	3	(7)	6	4	(10)	17
	S19	4	4	(8)	7	7	(14)	22
	S20	5	4	(9)	8	7	(15)	24
	Total	20	16	(36)	40	35	(75)	111
	Total	73	56	(129)	133	115	(248)	
								$G = 377$
								$N = 80$

represent the number of errors made by copy typists under various combinations of treatments (sleep deprivation, noise, article length and trials). Note that the hierarchically organized between-subject factors or levels (sleep deprivation and article length) are presented vertically, down the side of the matrix. The within-subject factors (noise and trials) are presented horizontally, across the top of the matrix. The five scores provided under each particular combination of treatments constitute the sixteen cells ($2 \times 4 + 2 \times 4$) and the total number of subjects $N = 16 \times 5 = 80$. A number of totals and sub-totals need to be calculated in a split-plot design and, as in Table 9.4, the matrix should allow room for the necessary totals to be inserted.

Following your construction of the appropriate data matrix, and before performing any actual ANOVA computations, a *summary table* needs to be drawn up. A summary table can also be thought of as a matrix, in which the results of your analysis can be clearly presented. Any summary table will contain six columns, but the number of rows will vary according to your particular design, and the number of factors investigated. Tables 9.6, 9.7 and 9.8 below represent the way in which the summary table should be set out for each of the three designs introduced earlier.

The rows of a summary table present the various possible *sources* of variation (i.e. each main effect, interaction and error) and the *total* variation obtained. In addition, within-subject and split-plot designs will include a between-subject source and a within-subject source. Both of these sources, as you will see later, represent sub-totals. The most difficult aspect of the organization of your summary table is

Table 9.6. Summary table for our example of a 2 between-subject (simple factorial) design

Source	*SS*	*DF*	*MS*	*F*-ratio	*P*
Sleep-deprivation (A)					
Article length (B)					
Interaction (A × B)					
Error					
Total					

Table 9.7. Summary table of our example of a 2 within-subject (repeated measures) design

Source	*SS*	*DF*	*MS*	*F-R*	*P*
(Between-subject		)			
(Within-subject		)			
Part 1:					
Noise (C)					
W-S error (*a*)					
Part 2:					
Trials (D)					
W-S error (*b*)					
Part 3:					
Interaction (C × D)					
W-S error (*c*)					
Total					

determining the number of error terms that you will need to calculate. For between-subject designs, there is only one error term for *all* the main effects and interaction effects. For within-subject designs and split-plot designs there can be more than one error term. The rule of thumb is that for repeated measures designs you need to calculate a separate error term for *each* main effect and *each* interaction effect. This also applies to the within-subject component of split-plot designs, but note from Table 9.7 that the component also contains the *mixed* interactions (i.e. those interactions involving both between-subject factors *and* within-subject factors). Thus each within-subject part of the summary table should not only contain the main within-subject effect, the interaction between the *main* effects and their respective error terms, but also the interactions of these effects with *each* of the between-subject main effects and interactions.

Once you have completed Module I, your data should be presented in a form directly amenable to the application of the relevant ANOVA computational procedures. Also, having drawn up the appropriate (incomplete) summary table, it serves as a guide to those 'sources' which need to be computed.

Table 9.8. Summary table for our 2 between-subject and 2 within-subject split-plot example

Source	*SS*	*DF*	*MS*	*F-R*	*P*
(Between-subjects		)			
Deprivation (A)					
Article length (B)					
Interaction (A × B)					
B-S Error					
(Within-subjects		)			
Part 1:					
Noise (C)					
Interaction (C × A)					
Interaction (C × B)					
Interaction (C × A × B)					
W-S.Error(*a*)					
Part 2:					
Trials (D)					
Interaction (D × A)					
Interaction (D × B)					
Interaction (D × A × B)					
W-S.Error(*b*)					
Part 3:					
Interaction (C × D)					
Interaction (C × D × A)					
Interaction (C × D × B)					
Interaction (C × D × A × B)					
W-S.Error(*c*)					
Total					

Module II: *Calculating the Total Sum of Squares*

This module applies to all ANOVA statistical designs.

The total sum of squares (TOT.SS) equals the total sum of squares from all sources. This total is important because not all the sources (i.e. the within-subject total and errors) can be calculated directly.

The formula for calculating the total sum of squares (TOT.SS) is the same for *all* ANOVA designs and is

(1) **TOT.SS** $= \Sigma X^2 - \dfrac{(\Sigma X)^2}{N}$

where ΣX^2 = sum of the squares of each observation (X), i.e. square first then add

$(\Sigma X)^2$ = the square of the sum of the observations, i.e. add first then square

N = the total number of observations

$\dfrac{(\Sigma X)^2}{N}$ = the correction term (CT)

To calculate the total number of degrees of freedom, the formula is

(2) **TOT.DF** $= N - 1$

In our split-plot example (Table 9.4),

$$\Sigma X^2 = 2^2 + 2^2 + 8^2 + \ldots + 8^2 + 7^2 = 2207$$

$$\text{CT} = \frac{(\Sigma X)^2}{N} = \frac{(2 + 2 + 8 + \ldots + 8 + 7)^2}{80} = 1776{\cdot}61$$

so that

$$\text{TOT.SS} = 2207 - 1776{\cdot}61 = 430{\cdot}39$$

with

$$\text{TOT.DF} = N - 1 = 80 - 1 = 79 \text{ DF}$$

Module III: *Calculating the Between-Subject Component*

The precise form of the between-subject component required depends on which of the three designs you are using.

In a *simple factorial* design, the between-subject sum of squares is equal to the total sum of squares, and so you will not need to carry out any additional computations.

In *repeated measures* designs you will need to calculate a separate between-subject sum of squares.

In *split-plot* designs you will need to calculate the between-subject sum of squares and the between-subject main effects, interaction effects and error term.

1 Calculation of the between-subject sum of squares (B-S.TOT) and degrees of freedom (B-S.DF)

Instead of squaring the individual observations as we did in calculating the total sum of squares, we square the various totals. The most complicated aspects of the calculation are finding the right totals to square, and finding the correct divisor (n). You should not confuse n with N (total number of observations) or k (number of levels). The n represents the number of observations that make up each total.

The formula for calculating the between-subject sum of squares (B-S.SS) is:

$$\text{(3) B-S.SS} = \frac{T_{s1}^2 + T_{s2}^2 + \ldots + T_{sn}^2}{n_s} - \text{CT}$$

where $T_{s1}^2 + T_{s2}^2 + \ldots + T_{sn}^2$ = the sum of the squared subject totals and n_s = the number of scores which made up each subject total.

To calculate the B-S.SS degrees of freedom (B-S.DF), the formula is

$$\text{(4) B-S.DF} = S - 1$$

where S = the number of subjects.

In our split-plot example,

$$\text{B-S.SS} = \frac{T_{S1}^2 + T_{S2}^2 + T_{S3}^2 + \ldots + T_{S19}^2 + T_{S20}^2}{n_S} - \text{CT}$$

so that

$$\text{B-S.SS} = \frac{20^2 + 18^2 + 18^2 + \ldots + 22^2 + 24^2}{4} - 1776{\cdot}61$$

$$= 1901{\cdot}25 - 1776{\cdot}61 = 124{\cdot}64$$

with

$$\text{B-S.DF} = 20 - 1 = 19 \text{ DF}$$

2 Calculation of the main effects

The formula for calculating the sum of squares of a treatment or main effect (ME(A).SS) is

$$\text{(5) ME(A).SS} = \frac{T_{A1}^2 + T_{A2}^2 + \ldots + T_{An}^2}{n_A} - \text{CT}$$

where $T_{A1}^2 + T_{A2}^2 + \ldots + T_{An}^2$ = the sum of the squares of the totals for each level in the main effect, and n_A = the number of observations which made up *each* of the above totals.

To calculate the ME(A) degrees of freedom (ME(A).DF) the formula is

(6) $\text{ME(A).DF} = k - 1$

where k = the number of levels in variable A.

In our split-plot example, there are two between-subject variables or main effects (A and B) each with two levels ($A1$, $A2$ and $B1$, $B2$). Thus:

$$\text{ME(A).SS} = \frac{T_{A1}^2 + T_{A2}^2}{n_A} - \text{CT and ME(B).SS} = \frac{T_{B1}^2 + T_{B2}^2}{n_B} - \text{C}$$

where

$$T_{A1}^2 = (91 + 61)^2 = 152^2 \text{ and } T_{B1}^2 = (91 + 114)^2 = 205^2$$
$$T_{A2}^2 = (114 + 111)^2 = 225^2 \qquad T_{B2}^2 = (61 + 111)^2 = 172^2$$
$$n_A = 40 \qquad n_B = 40$$

so that

$$\text{ME(A).SS} = \frac{23104 + 50625}{40} - 1776{\cdot}61 = 66{\cdot}61$$

and

$$\text{ME(B).SS} = \frac{42025 + 29584}{40} - 1776{\cdot}61 = 13{\cdot}61$$

With respect to calculating the degrees of freedom, use formula 6 where $k = 2$ for both effects, so that:

$$\text{ME(A).DF} = 2 - 1 = 1 \text{ DF and ME(B).DF} = 2 - 1 = 1 \text{ DF.}$$

Note that the calculation of the between-subject sum of squares outlined in the last section is essentially the same as a calculation of a main effect. 'Subjects' can be thought of as a factor (or main effect) with each subject representing one of the levels of the factor. Therefore, in the between-subject calculation n_S is used in the same way as n_A is in the case of the calculation of a main effect, and $S - 1$ is the equivalent of $k - 1$ in the respective calculations of the degrees of freedom.

3 Calculation of the interaction effect(s)

The calculation of interaction effects also involves totals. However, selecting the right totals with which to calculate the various interaction effects is not always easy. In order to help you find the correct totals to square, we would suggest that you draw up an *interaction table* for every interaction except the final interaction. (This is because the final interaction totals are easy to determine since they will equal the cell totals.) Figures 9.5 to 9.12 represent some examples of interaction tables, along with their graphic illustrations.

The formula for calculating the sum of squares for an interaction (INT(A × N) is

$$\text{(7) INT(A} \times \text{N).SS} = \frac{T_{A1N1}^2 + T_{A1N2}^2 + \ldots + T_{AnNn}^2}{n_{AN}} - (\text{CT} + \text{ME(A} \ldots \text{N).SS} + \text{INT(A} \times \ldots \text{N).SS})$$

where:

$T_{A1N1}^2 + T_{A1N2}^2 + \ldots + T_{AnNn}^2$ = the sum of the squared interaction totals for all combinations of A to N, at all levels.

ME(A ... N).SS = all the main effects subsumed within the interaction. For example, a two-way interaction (A × B) has two main effects (A and B) subsumed. A three-way interaction (A × B × C) has three main effects (A, B, C) subsumed, and so on.

INT(A × . . .N).SS = all the interaction effects subsumed within the interaction being calculated. For example, a two-way interaction has no other interactions subsumed. A three-way interaction (A × B × C) has three two-way interactions subsumed (A × B; A × C; B × C). A four-way interaction (A × B × C × D) has six two-way interactions (A × B, A × C, A × D, B × C, B × D and C × D) and four three-way interactions (A × B × C, A × B × D, A × C × D and B × C × D).

n_{AN} = the number of observations which make up each of the interaction totals,

CT = correction term.

To calculate the interaction degrees of freedom (INT(A × B).DF) use the formula

(8) INT(A × N).DF = ME(A).DF × . . . ME(N).DF

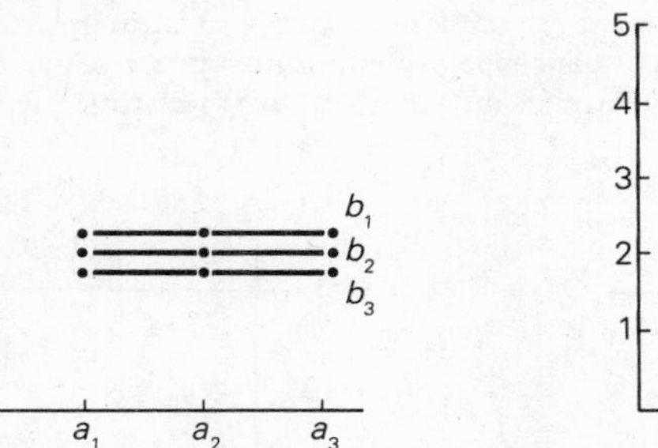

		Treatment a			
		a_1	a_2	a_3	T_b
Treatment b	b_1	2	2	2	6
	b_2	2	2	2	6
	b_3	2	2	2	6
	T_a	6	6	6	

Figure 9.5

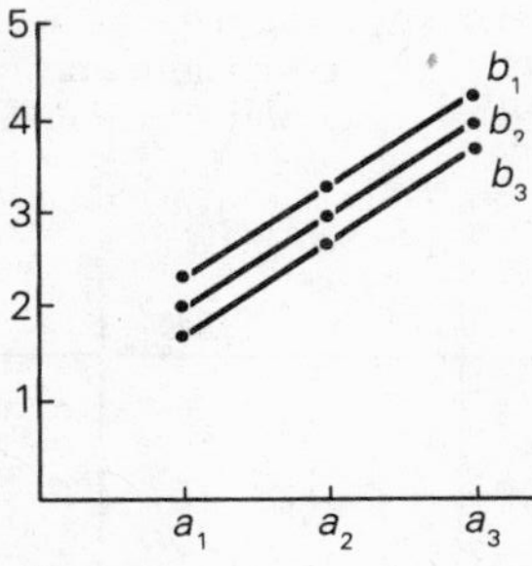

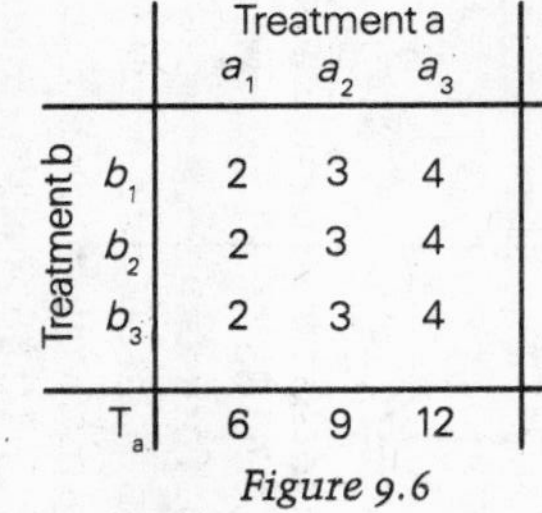

		Treatment a			
		a_1	a_2	a_3	T_b
Treatment b	b_1	2	3	4	9
	b_2	2	3	4	9
	b_3	2	3	4	9
	T_a	6	9	12	

Figure 9.6

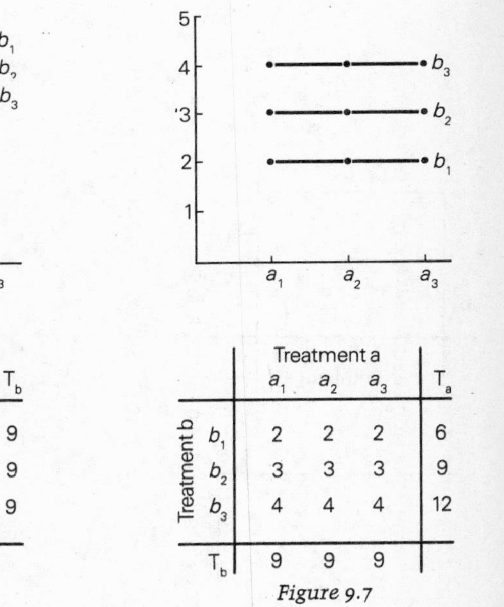

		Treatment a			
		a_1	a_2	a_3	T_a
Treatment b	b_1	2	2	2	6
	b_2	3	3	3	9
	b_3	4	4	4	12
	T_b	9	9	9	

Figure 9.7

Figure 9.5 No significant main effects or interaction
Figure 9.6 Significant A effect; non-significant B effect and interaction
Figure 9.7 Significant B effect; non-significant A effect and interaction

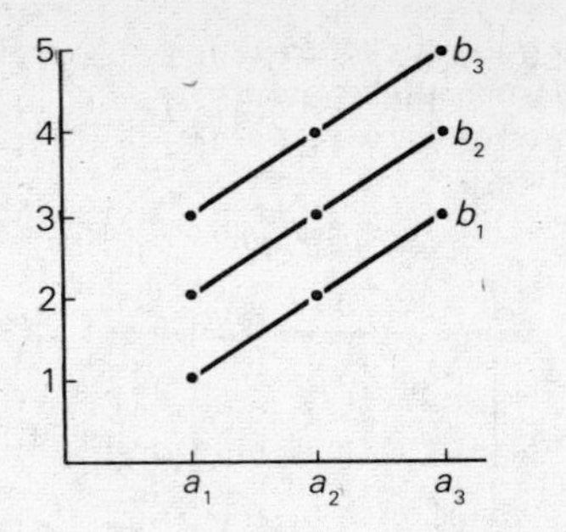

Treatment b	Treatment a a_1	a_2	a_3	T_b
b_1	1	2	3	6
b_2	2	3	4	9
b_3	3	4	5	12
T_a	6	9	12	

Figure 9.8

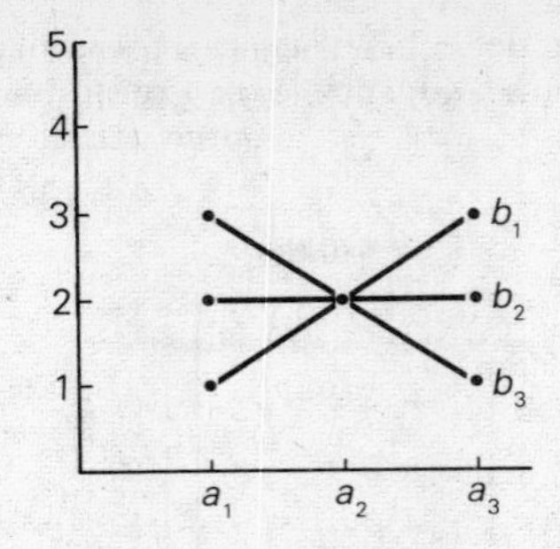

Treatment b	Treatment a a_1	a_2	a_3	T_b
b_1	1	2	3	6
b_2	2	2	2	6
b_3	3	2	1	6
T_a	6	6	6	

Figure 9.9

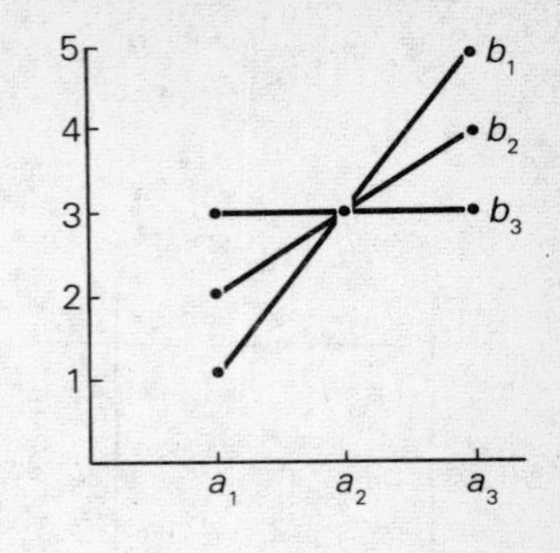

Treatment b	Treatment a a_1	a_2	a_3	T_b
b_1	1	3	5	9
b_2	2	3	4	9
b_3	3	3	3	9
T_a	6	9	12	

Figure 9.10

Figure 9.8 Significant A and B effects; no significant interaction
Figure 9.9 Non-significant A and B effects; significant interaction
Figure 9.10 Significant A effect and interaction but non-significant B effect

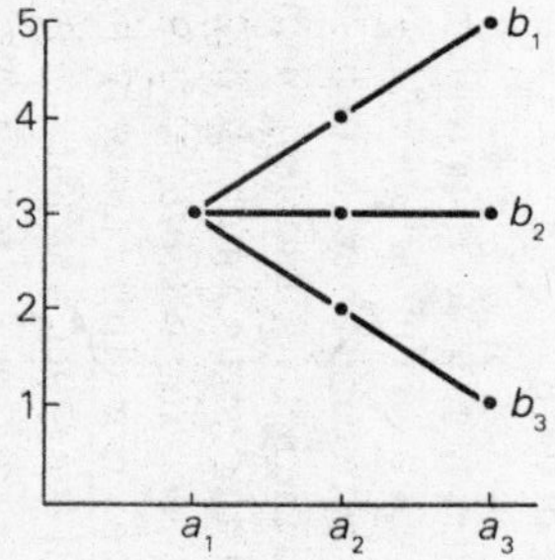

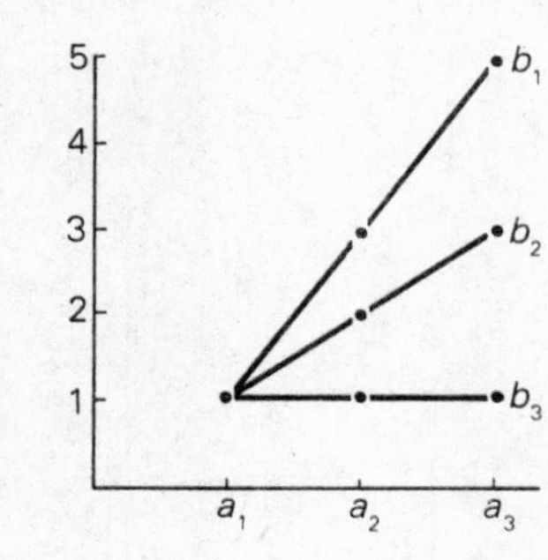

Treatment b	Treatment a a_1	a_2	a_3	T_b
b_1	3	4	5	12
b_2	3	3	3	9
b_3	3	2	1	6
T_a	9	9	9	

Figure 9.11

Treatment b	Treatment a a_1	a_2	a_3	T_b
b_1	1	3	5	9
b_2	1	2	3	6
b_3	1	1	1	3
T_a	3	6	9	

Figure 9.12

Figure 9.11 Significant B effect and interaction but non-significant A effect
Figure 9.12 All main effects (A and B) and interaction are significant

In our split-plot example, there is only one between-subject interaction (A × B).

First, draw up an interaction table of totals:

	A1	*A2*
B1	91	114
B2	61	111

Secondly, calculate the A × B interaction sum of squares by using formula 7:

$$\text{INT(A} \times \text{B).SS} = \frac{T_{A1B1}{}^2 + T_{A1B2}{}^2 + T_{A2B1}{}^2 + T_{A2B2}{}^2}{n_{\text{AB}}}$$

$$- (\text{CT} + \text{ME(A).SS} + \text{ME(B).SS})$$

where

$$T_{A1B1}{}^2 = 91^2 = 8281$$
$$T_{A1B2}{}^2 = 61^2 = 3721$$
$$T_{A2B1}{}^2 = 114^2 = 12996$$
$$T_{A2B2}{}^2 = 111^2 = 12321$$
$$n_{AB} = 20$$

so that

$$\text{INT(A} \times \text{B).SS} = \frac{8281 + 3721 + 12996 + 12321}{20}$$
$$- (1776{\cdot}61 + 66{\cdot}61 + 13{\cdot}61)$$
$$= 1865{\cdot}95 - 1856{\cdot}83 = 9{\cdot}12$$

Calculate the degrees of freedom using formula 8:

$$\text{A} \times \text{B.DF} = \text{ME(A).DF} \times \text{ME(B).DF} = 1 \times 1 = 1$$

4 Calculation of the between-subject error term

All error terms are calculated by residual, and as usual the problem is to find the correct total from which to subtract the various effects (main and interaction). In the case of the between-subject error, it is the between-subject total (B-S.TOT) and the between-subject main and interaction effects that are required. In designs which are only between-subject, if we recall that *the between-subject total is equal to the total sums of squares*, then the error term can only be determined after you have calculated all the other relevant totals, main effects and interactions. The degrees of freedom for the error term are also calculated by residual, that is by subtracting all the other degrees of freedom from the total number of degrees of freedom.

The formula for calculating the sum of squares for the between-subject error (B.ERR.SS) is

(9) B.ERR.SS = B-S.TOT − (ME(A . . . N).SS + INT(A × . . . N).SS)

where

B-S.TOT = the between-subject total
ME(A . . . N).SS = all the between-subject main effects
INT(A × . . . N).SS = all the between-subject interactions

To calculate the B.ERR.SS degrees of freedom (B.ERR.DF) use

(10) B.ERR.DF = B-S.DF − (ME(A . . . N).DF + INT(A × . . . N).DF)

In our split-plot worked example

B.ERR.SS = B-S.SS − (ME(A).SS + ME(B).SS + INT(A × B).SS)

where

B-S.SS = 124·63; ME(A).SS = 66·61
ME(B).SS = 13·61; INT(A × B).SS = 9·11

so that

B.ERR.SS = 124·63 − 89·33 = 35·30

For calculating degrees of freedom use formula 10, so that

B.ERR.DF = 19 − (1 + 1 + 1) = 19 − 3 = 16 DF

Module IV: *Calculation of the Within-Subject Component*

This component is *required only for repeated measures and split-plot designs.* For both designs the initial step will be to calculate the within-subject sum of squares. The number of further parts required to make up the complete calculation depends on the size of your design.

In *repeated measures* designs, you will need to compute separate parts for each individual *main effect* and *interaction effect*. Each part will also contain the error term for each effect. For example, in a two-way (C by D) repeated measures design the within-subject component will contain three parts in addition to the within-subject sum of squares. These are part 1, main effect C plus error (*a*); part 2, main effect D plus error (*b*); and part 3, interaction C × D plus error (*c*) (see summary Table 9.7).

In *split-plot* designs there are no *extra parts* to compute, but each part must incorporate the relevant within-by-between variable interaction. For example, in a one between (C) by two within (A by B) split-plot, the within-subject component will contain part 1, main effect A, interaction A × C plus error term (*a*); part 2, main effect B, interaction B × C plus error term (*b*); and part 3, interactions A × B and A × B × C plus error term (*c*).

1 Calculation of the within-subject sum of squares

The within-subject sum of squares is quite simple to calculate. The total sum of squares in split-plot and repeated measures designs consists of two parts: the between-subject sum of squares (B-S.SS) and the within-subject sum of squares (W-S.SS). Having already calculated the total sum of squares (in Module II) and the between-subject sum of squares (in Module III), then the within-subject sum of squares can be calculated by residual.

The formula for calculating the within-subject sum of squares (W-S.SS) is thus

(11) W-S.SS = TOT.SS − B-S.SS

where TOT.SS = the total sum of squares
B-S.SS = the between-subject sums of squares.

The formula to calculate the W-S.SS degrees of freedom (W-S.DF) is

(12) W-S.DF = TOT.DF − B-S.DF

In our split-plot example,

W-S.SS = TOT.SS − B-S.SS

where

TOT.SS = 430·39
B-S.SS = 124·64

so that

W-S.SS = 430·39 − 124·64 = 305·75

The degrees of freedom can be calculated by formula 12:

W-S.DF = 79 − 19 = 60

2 Calculation of the error terms

Calculating the within-subject error terms involves some new computational procedures. The calculation of these error terms is not in itself difficult, but with more than one error term, confusion can arise. There are some rules of thumb which should help.

(*i*) Recall that there is an error term for each within-subject main effect and interaction effect (e.g. there is one error (*a*) for C, one (*b*) for D, and one (*c*) for C × D in our split-plot design).

(*ii*) Like all error terms, they are calculated by residual. The total for each error term is treated as if it was an interaction between the main effect and the subject effect (e.g. C × S or D × S or C × D × S). The subsumed effects are then subtracted, leaving only the error term. The exception to this rule is the calculation of the final error term, which is calculated by subtracting all the within-subject effects from the within-subject total. The remainder then equals the final error term.

Calculation of all but the final error term. Before you can calculate your error terms, you need to calculate the main effect or interaction SS and DFs associated with the particular error term. These are calculated in the normal way using formulae 5, 6, 7 and 8. Do these yourself and check your answers with the SS and DFs given in Table 9.9.

The final calculation involved in this first *part* is the error term (*a*). It is not the last error term, so you use the following formula:

(13a) W-S.ERR.SS(*a*) $= \dfrac{(\text{ME(A)/INT(A} \times \text{N).SS} \times \text{S)}}{n\text{ME(A)/INT(A} \times \text{N)}}$ **− (CT + ME(A . . . N)/INT(A × . . . N).SS + B-S.SS)**

And to calculate the error degrees of freedom (W-S.ERR.DF(*a*)) use:

(14a) W-S.ERR.DF(*a*) = ME(A)/INT(A × N).DF × S − ME(A . . .N)/INT(A × . . . N).DF

In our split-plot example the first within-subject error is error (*a*), so that

$$\text{W-S.ERR.SS}(a) = \frac{\text{C} \times \text{S}}{n_{CS}} - (\text{CT} + \text{ME(C).SS} + \text{INT(C} \times \text{A). SS} + \text{INT (C} \times \text{B).SS} + \text{INT (C} \times \text{A} \times \text{B). SS} + \text{B-S.SS})$$

where

C × S = $T_{C1S1}{}^2 + T_{C1S2}{}^2 + \ldots + T_{CnSn}{}^2$
ME(A).SS = the sum of squares of main effect C
INT(C × A).SS = the sum of squares for the C × A interaction
INT (C × B).SS = the sum of squares for the C × B interaction
INT(C × A × B).SS = the sum of squares for the C × A × B interaction
B-S.SS = the between-subject sum of squares
n_{CS} = number of observations making up each total

Draw up on the appropriate interaction table:

	S1	*S2*	*S3*	*S4* . . .	*S19*	*S20*
C1	4	2	5	3 . . .	8	9
C2	16	16	13	15 . . .	14	15

Therefore:

$$\text{W-S.ERR.SS}(a) = \frac{4^2 + 2^2 + 5^2 + 3^2 \ldots 14^2 + 15^2}{2} - (1776{\cdot}61 + 177{\cdot}01 + 0{\cdot}61 + 15{\cdot}31 + 52{\cdot}81 + 124{\cdot}63)$$

$$= \frac{4361}{2} - 2146{\cdot}98 = 33{\cdot}52$$

To calculate the degrees of freedom use formula 14*a*:

$$\text{W-S.ERR.DF}(a) = \text{ME(C).DF} \times \text{S} - (\text{ME(C).DF} + \text{INT(C} \times \text{A).DF} + \text{INT(C} \times \text{B).DF} + \text{INT(C} \times \text{A} \times \text{B).DF})$$

so that

W-S.ERR.DF(a) = 1 × 19 − 4 = 16

In our split-plot example, we have also to calculate the error term for the second *part* (the D main effect). This involves repeating the above procedures and formulae, so that for error term (b):

W-S.ERR.SS(b) = 4·7 with 16 DF

The final error term. The final (or *the first if the only one*) within-subject error term is calculated by using formula 13b below:

(13b) W-S.ERR.SS(2) = W-S.SS − all other within-subject SS

Likewise the formula for the degrees of freedom is similar:

(14b) W-S.ERR.DF(2) = W-S.DF − all other within-subject DFs

In our split-plot example:

W-S.ERR.SS(c) = W-S.SS − all other SS

so that

W-S.ERR.SS(c) = 305·75 − 300·84 = 4·9

To calculate the degrees of freedom:

W-S.ERR.DF(c) = W-S.DF − all other within-subject DFs

so that

W-S.ERR.DF(c) = 60 − 44 = 16

Module V: *The Calculation of the Mean Squares and F-ratios*

We now need to calculate the mean squares (MS). These are calculated by dividing the main and interaction effects by their respective degrees of freedom. (These formulae apply to *all* ANOVA designs.) Therefore:

$$(15)\ \mathbf{ME(A).MS} = \frac{\mathbf{ME(A).SS}}{\mathbf{ME(A).DF}}\ \textbf{and}\ \mathbf{INT(A \times N).MS} = \frac{\mathbf{INT(A \times N).S}}{\mathbf{INT(A \times N).D}}$$

The F-ratio is then calculated by dividing each mean square by the mean square of the appropriate error term. (F-ratios are usually calculated only for main effects or interaction effects.) Therefore:

$$(16)\ \text{F-R(A)} = \frac{\text{ME(A).MS}}{\text{ERROR.MS}} \quad \text{or} \quad \text{F-R(A} \times \text{N)} = \frac{\text{INT(A} \times \text{N).MS}}{\text{ERROR.MS}}$$

In our split-plot example, there is one error term for the between-subject component and three in the within-subject component.

When calculating your mean squares and *F*-ratios, make use of your partially completed summary table, and enter the results of the calculations under the appropriate headings (see Table 9.9).

The probabilities that should be entered under the column headed '*P*' are determined by looking up the *F*-ratios in the *F*-ratio tables (Table IV in the Appendix). Two sets of degrees of freedom are required, along with each ratio. These are the DF for the particular effect (A, B, A × B etc.) and the DF for the relevant error term. For example, to determine the significance of the *F*-ratio for ME(A) in our split-plot example you would look up the probability associated with an *F*-ratio of 30·19 with 1 and 16 DFs. The *F*-ratio at a probability level of $P = 0{\cdot}001$ is 16·12. As our *F*-ratio is considerably larger than this value we can state that the probability of this value being obtained by chance is less than 0·001 ($P < 0{\cdot}001$). Therefore, the larger the *F*-ratio, and the higher the two sets of degrees of freedom,

Table 9.9. Summary table illustrating the mean squares and F-ratio calculations in the split-plot example

Source	SS	DF	MS	F-R	*P*
(Between-subjects	124·64	19)			
ME(A)	66·61	1	$\frac{66{\cdot}61}{1} = 66{\cdot}61$	$\frac{66{\cdot}61}{2{\cdot}21} = 30{\cdot}19$	$P < 0{\cdot}001$
ME(B)	13·61	1	$\frac{13{\cdot}61}{1} = 13{\cdot}61$	$\frac{13{\cdot}61}{2{\cdot}21} = 6{\cdot}17$	$P < 0{\cdot}05$
INT(A × B)	9·12	1	$\frac{9{\cdot}12}{1} = 9{\cdot}12$	$\frac{9{\cdot}12}{2{\cdot}21} = 4{\cdot}13$	NS
B-S Error	35·30	16	$\frac{35{\cdot}30}{16} = 2{\cdot}21$		
(Within-subjects	305·75	60)			

Table 9.9. – cont.

Source	SS	DF	MS	F-R	*P*
Part 1:					
ME(C)	177·01	1	$\frac{177.01}{1} = 177.01$	$\frac{177.01}{2.09} = 84.69$	$P < 0.001$
INT(C × A)	0·61	1	$\frac{0.61}{1} = 0.61$	$\frac{0.61}{2.09} = 0.29$	NS
INT(C × B)	15·31	1	$\frac{15.31}{1} = 15.31$	$\frac{15.31}{2.09} = 7.32$	$P < 0.05$
INT(C × A × B)	52·81	1	$\frac{52.81}{1} = 52.81$	$\frac{52.81}{2.09} = 25.27$	$P < 0.001$
W-S.Error(*a*)	33·52	16	$\frac{33.52}{16} = 2.09$		
Part 2:					
ME(D)	15·31	1	$\frac{15.31}{1} = 15.31$	$\frac{15.31}{0.29} = 52.79$	$P < 0.001$
INT(D × A)	0·11	1	$\frac{0.11}{1} = 0.11$	$\frac{0.11}{0.29} = 0.38$	NS
INT(D × B)	0·11	1	$\frac{0.11}{1} = 0.11$	$\frac{0.11}{0.29} = 0.38$	NS
INT(D × A × B)	0·01	1	$\frac{0.01}{1} = 0.01$	$\frac{0.01}{0.29} = 0.03$	NS
W-S.Error(*b*)	4·70	16	$\frac{4.70}{16} = 0.29$		
Part 3:					
INT(C × D)	0·01	1	$\frac{0.01}{1} = 0.01$	$\frac{0.01}{0.31} = 0.03$	NS
INT(C × D × A)	0·61	1	$\frac{0.61}{1} = 0.61$	$\frac{0.61}{0.31} = 1.97$	NS
INT(C × D × B)	0·11	1	$\frac{0.11}{1} = 0.11$	$\frac{0.11}{0.31} = 0.35$	NS
INT(C × D × A × B)	0·61	1	$\frac{0.61}{1} = 0.61$	$\frac{0.61}{0.31} = 1.97$	NS
W-S.Error(*c*)	4·90	16	$\frac{4.90}{16} = 0.31$		
Total	430·39	79			

the more likely it is that your F-ratio will be significant (i.e. the smaller the F value needs to be to reach the critical value).*

Some useful computational checks

(a) *It is not possible to have a negative sum of squares.* If you do obtain a negative sum of squares then you must have made an error in the calculations. Usually errors occur because the wrong totals have been squared or incorrectly summed, or because the correction term is wrong.

(b) Note that *all the levels of any main effect, and all the combination of levels in any interaction effect, must total the grand total* (ΣX). For example, when calculating the main effect A, check that $T_{A1} + T_{A2} = \Sigma X$. For the interaction A × B check that $T_{A1B1} + T_{A1B2} + T_{A2B1} + T_{A2B2} = \Sigma X$. If they do not, your sub-totals or grand total may be wrong. With two-way interaction tables you can often check the totals by summing the rows and columns. These sums should equal the two main effect totals which constitute that interaction. The interaction tables in Figures 9.5 to 9.12 provide some examples of such totals.

(c) When discussing the computation of the between-subject sum of squares, we indicated a potential problem of confusing n with k. This also applies to the calculation of the main and interaction sums of squares. Recall that n is the number of observations that make up each total, and k is the number of levels. *To check you have the right n or k note that $n \times k = N$* (where N is the total number of observations). In our main effect A, we had two levels ($k = 2$), a total of 80 observations ($N = 80$), and $n = 40$ so that $N = n \times k$ ($80 = 40 \times 2$). This can also be applied to interactions. For example our A × B interaction had two levels in both factors ($k_1 = 2, k_2 = 2$), $N = 80$, and $n = 20$. Thus $N = k_1 \times k_2 \times n$ or $80 = 2 \times 2 \times 20$. However, it is unwise to use this equation for actually determining n (e.g. $n = N/k$). It is essentially *only* a checking procedure.

* There are some misunderstandings about whether F-ratio tables give critical values for a one-tailed or two-tailed test. F-ratio tables are one-tailed in a *limited* sense but *not* in terms of predicting the direction of your finding (i.e. A < B or A > B); in these circumstances the tables are two-tailed. Therefore, *if you predict the direction of an effect* then you would be justified in halving the associated probabilities given in the tables. Under such circumstances the 10 per cent ($P = 0{\cdot}1$) values in the tables would be equivalent to a one-tailed 5 per cent ($P = 0{\cdot}05$) value.

(*d*) In your results *the larger F-ratios should tally with the larger differences between means.* If they don't something is wrong.

Module VI: *Results and Interpretation*

This final module, like the first module, involves no calculations. It is concerned instead with the interpretation of your results.

You should examine and interpret your means in the light of your hypotheses and significant levels. You should do this *in prose,* with the help of tables where you think they may be useful. Tables and graphs are particularly useful when dealing with interactions. There are no hard and fast rules but if the means can be placed in the text then they should be. If, however, there are too many means, or if you are dealing with interactions, then tables or graphs should ideally be used.

Locate the critical means and assess whether the differences lie in the direction you predicted. Then check whether the effect is significant. Do *not* base your discussion on the significant levels without looking at the means – they could differ in the opposite direction to your prediction! Having interpreted your main and interaction effects by relating the results obtained to your hypotheses, you should then attempt to interpret any remaining, but not necessarily predicted, significant findings. Such findings should be treated with care, and should not be used to make *post facto* rationalizations in the form of changing old hypotheses or generating new ones (see also p. 31). However, their explorative use is often most valuable, as they can provide the basis for further hypotheses and thus future experimentation.

In summary:

(*a*) Have the tables of means and summary table at hand.
(*b*) Select and interpret those means for which you predicted effects on the basis of your hypotheses.
(*c*) Interpret the means of any unpredicted though significant effects (but with a degree of caution).

In order to demonstrate how you might interpret the results of a particular investigation, we will return to our two-way simple factorial example.

Interpretation of our two-way simple factorial ANOVA example (see Table 9.3)

(a) Possible predictions

(i) That sleep-deprived subjects will make more typing errors than non-sleep-deprived subjects.
(ii) That the number of typing errors will increase as the length of the article to be typed increases.
(iii) That this increase in errors will be most marked for the sleep-deprived subjects.

The ANOVA design adopted is between-subject, with two main effects (sleep deprivation and article length), an interaction effect (sleep deprivation by article length) and an error effect.

(b) The means and summary table

Means:

Main effects:
Sleep-deprived ($A1$) $\bar{X} = 4.6$
Non-sleep-deprived ($A2$) $\bar{X} = 3.0$

Article 1 ($B1$) $\bar{X} = 1.4$
Article 2 ($B2$) $\bar{X} = 3.8$
Article 3 ($B3$) $\bar{X} = 6.2$

Sleep deprivation by article length (A × B) Interaction:

Interaction Table of Means:

	*B*1	*B*2	*B*3
*A*1	1·6	4·4	7·8
*A*2	1·2	3·2	4·6

Graphic Representation:

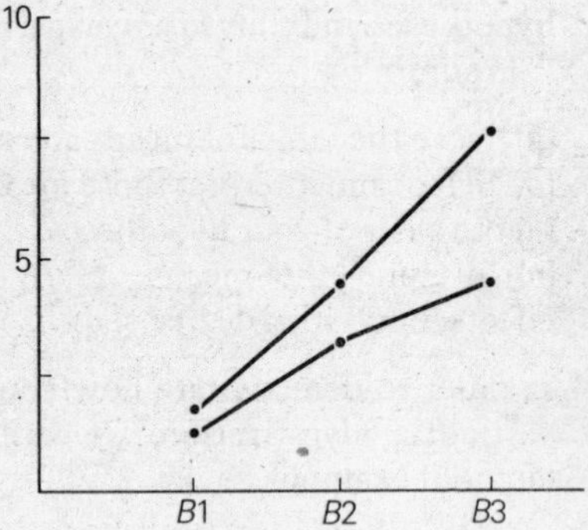

For the ANOVA summary table see Table 9.11(ii).

(*c*) Interpretation

As predicted, sleep-deprived subjects ($\bar{x}$ = 4·6) produced significantly more typing errors than the controls ($\bar{x}$ = 3·0: F = 32·93, DF = 1,24, $P < 0{\cdot}001$). In addition, errors significantly increased with article length (F = 99·3, DF = 2,24, $P < 0{\cdot}001$). More importantly, the increase in errors was greater for the sleep-deprived group than for the control group (see figure opposite). This interaction was significant (F = 8·96, DF = 2,24, $P < 0{\cdot}01$). Therefore, the prediction that fatigue effects have a more rapid onset when subjects are sleep-deprived has been supported.

The examples in modular form

In this chapter, of the three designs for which we presented hypothetical data matrices only our split-plot design has been fully worked through in modular form. However, it is hoped that the modular approach, which has presented the ANOVA steps in a summarized form, will help you with both computer and hand computations. If you are using a computer then Modules I and VI are obviously the critical ones.

Tables 9.10, 9.12 and 9.14 show the three basic ANOVA designs

Table 9.10. Step-by-step procedures in modular form for our example of a two-way between-subject (simple factorial) design

Step	Module	Source	Formula
Step 1	Module I	(Data presentation)	(none)
Step 2	Module II	TOT.SS/DF	(1)/(2)
Step 3	Module III	ME(A).SS/DF	(3)/(4)
Step 4		ME(B).SS/DF	(3)/(4)
Step 5		INT(A × B).SS/DF	(5)/(6)
Step 6		B.ERR.SS/DF	(7)/(8)
Step 7	Module V	MS/F-R	(15)/(16)
Step 8	Module VI	(Interpretation)	(none)

Table 9.11. Step-by-step worked example of our two-way between-subject (simple factorial) design

Step 1 Module I: Data presentation

(i) The factorial matrix:

Article length (*B*)	Sleep-deprived (*A*1)			Non-sleep-deprived (*A*2)		
	*B*1	*B*2	*B*3	*B*1	*B*2	*B*3
	1	4	9	1	2	5
	3	4	8	0	4	4
	2	5	8	1	3	4
	1	5	7	2	4	5
	1	4	7	2	3	5
Total	8	22	39	6	16	23

Grand total (G) = 114
N = 30

(ii) Summary table:

Source	SS	DF	MS	*F*-ratio	*P*
Sleep deprivation (*A*)	19·2	1	19·2	32·93	$P < 0{\cdot}001$
Article length (*B*)	115·2	2	57·63	99·31	$P < 0{\cdot}001$
Interaction (*A* × *B*)	10·4	2	5·2	8·96	$P < 0{\cdot}01$
Error	14·0	24	0·58		
Total	158·8	29			

Step 2 Module II: Calculate total SS/DF

(1) Formula (1): **TOT.SS** $= \Sigma X^2 - \dfrac{(\Sigma X)^2}{N} = 1^2 + 4^2 + 9 \ldots 5^2 - \dfrac{(114)^2}{30}$

$= 592 - 433{\cdot}2 = 158{\cdot}8$

(ii) Formula (2): **TOT.DF** $= N - 1 = 30 - 1 = 29$ DF.

Module III: Calculate the between-subject component

Table 9.11. – cont.

Step 3

(i) Formula (5): $\mathbf{ME(A).SS} = \dfrac{T_{A1}{}^2 + T_{A2}{}^2}{n_A} - \mathbf{CT} = \dfrac{69^2 + 45^2}{15} - 433{\cdot}2$

$= 452{\cdot}4 - 433{\cdot}2 = 19{\cdot}2$

(ii) Formula (6): $\mathbf{ME(A).DF} = k - 1 = 2 - 1 = 1$ DF.

Step 4

(i) Formula (5): $\mathbf{ME(B).SS} = \dfrac{T_{B1}{}^2 + T_{B2}{}^2 + T_{B3}{}^2}{n_B} - \mathbf{CT}$

$= \dfrac{14^2 + 38^2 + 62^2}{10} - 433{\cdot}2$

$= 548{\cdot}4 - 433{\cdot}2 = 115{\cdot}2$

(ii) Formula (6): $\mathbf{ME(B).DF} = k - 1 = 3 - 1 = 2$ DF.

Step 5

(i) Formula (7): $\mathbf{INT(A \times B).SS} = \dfrac{T_{A1B1}{}^2 + T_{A1B2}{}^2 + \ldots + T_{A2B3}{}^2}{n_{AB}}$

$- (\text{CT} + \text{ME(A).SS} + \text{ME(B).SS})$

$= \dfrac{8^2 + 22^2 + \ldots + 23^2}{5}$

$- (433{\cdot}2 + 19{\cdot}2 + 115{\cdot}2)$

$= 578 - 567{\cdot}6 = 10{\cdot}4$

(ii) Formula (8): $\mathbf{INT(A \times B).DF} = \mathbf{ME(A).DF \times ME(B).DF}$

$= 1 \times 2 = 2$ DF.

Step 6

(i) Formula (9): $\mathbf{B\text{-}S.ERR.SS} = \mathbf{TOT.SS - (ME(C).SS + ME(D).SS + INT(C \times D).SS)}$

$= 158{\cdot}8 - (19{\cdot}2 + 115{\cdot}2 + 10{\cdot}4) = 14{\cdot}0$

(ii) Formula (10): $\mathbf{B\text{-}S.ERR.DF} = \mathbf{TOT.DF - (ME(C).DF + ME(D).DF + INT(C \times D).DF)}$

$= 29 - (1 + 2 + 2) = 24$ DF.

Step 7 Module V: Calculate mean squares and F-ratios

(i) $\mathbf{MS(C)} = \dfrac{\mathbf{ME(C).SS}}{\mathbf{ME(C).DF}} = \dfrac{19{\cdot}2}{1} = 19{\cdot}2$

(ii) $\mathbf{MS(D)} = \dfrac{\mathbf{ME(D).SS}}{\mathbf{ME(D).DF}} = \dfrac{115{\cdot}2}{2} = 57{\cdot}6$

Table 9.11. – cont.

$$\text{(iii)}\ \mathbf{MS(C \times D)} = \frac{\mathbf{INT(C \times D).SS}}{\mathbf{INT(C \times D).DF}} = \frac{10{\cdot}4}{2} = 5{\cdot}2$$

$$\text{(iv)}\ \mathbf{MS(ERR)} = \frac{\mathbf{B\text{-}S.ERR.SS}}{\mathbf{B\text{-}S.ERR.DF}} = \frac{14{\cdot}0}{24} = 0{\cdot}58$$

$$\text{(v)}\ \mathbf{F\text{-}R(C)} = \frac{\mathbf{MS(C)}}{\mathbf{MS(ERR)}} = \frac{19{\cdot}1}{0{\cdot}58} = 32{\cdot}93$$

$$\text{(vi)}\ \mathbf{F\text{-}R(D)} = \frac{\mathbf{MS(D)}}{\mathbf{MS(ERR)}} = \frac{57{\cdot}6}{0{\cdot}58} = 99{\cdot}31$$

$$\text{(vii)}\ \mathbf{F\text{-}R(C \times D)} = \frac{\mathbf{MS(C \times D)}}{\mathbf{MS(ERR)}} = \frac{5{\cdot}2}{0{\cdot}58} = 8{\cdot}96$$

Step 8   Module VI: Interpretation (see pp. 179–80)

---

presented in modular form. With the first two examples (simple factorial and repeated measures) we have also provided the equivalent worked example, based on the data contained in Tables 9.3 and 9.4.

Table 9.12. A two-way repeated measures ANOVA in modular form

Step	**Module**	**Source**	**Formula**
Step 1	Module I	(Data presentation)	(none)
Step 2	Module II	TOT.SS/DF	(1)/(2)
Step 3	Module III	B-S.SS/DF	(9)/(10)
Step 4	Module IV	W-S.SS/DF	(11)/(12)
Step 5	Part 1	ME(C).SS/DF	(3)/(4)
Step 6		W-S.ERR(*a*)	(13*a*)/(14*a*)
Step 7	Part 2	ME(D).SS/DF	(3)/(4)
Step 8		W-S.ERR(*b*)	(13*a*)/(14*a*)
Step 9	Part 3	INT(C × D).SS/DF	(5)/(6)
Step 10		W-S.ERR(*c*)	(13*b*)/(14*b*)
Step 11	Module V	MS/F-R	(15)/(16)
Step 12	Module VI	(interpretation)	(none)

Table 9.13. A worked example of our two-way within-subject (repeated measure) design

Step 1 Module I: Data presentation

(i) Data matrix:

Trials (*D*)	60 dbs (*C1*) *D1*	*D2*	*D3*	*D4*	85 dbs (*C2*) *D1*	*D2*	*D3*	*D4*	S.T.
S1	2	4	5	6	1	4	6	9	37
S2	4	5	6	7	2	4	7	10	45
S3	3	3	6	8	3	3	7	8	41
S4	2	4	5	8	3	5	8	10	45
S5	2	3	7	8	2	5	7	10	44
Total	13	19	29	37	11	21	35	47	$G = 212$

(ii) Summary Table:

Source	SS	DF	MS	F-R	*P*
(Between-subject	5·9	4)			
(Within-subject	240·5	35)			
Noise (*C*)	6·4	1	6·4	8·31	$P < 0{\cdot}05$
W-S.ERR(*a*)	3·1	4	0·77		
Trials (*D*)	209·2	3	69·73	134·1	$P < 0{\cdot}001$
W-S.ERR(*b*)	6·3	12	0·52		
Interaction ($C \times D$)	8·0	3	2·67	4·3	$P < 0{\cdot}05$
W-S.ERR(*c*)	7·5	12	0·62		
Total	246·4	39			

Step 2 Module II: Calculate the total SS/DF

(i) Formula (1): **TOT.SS** $= \Sigma X^2 - \dfrac{(\Sigma X)^2}{N} = 2^2 + 4^2 + 5^2 + \ldots + 10^2 - \dfrac{(212)^2}{40}$

$= 1370 - 1123{\cdot}6 = 246{\cdot}4$

Table 9.13. – cont.

(ii) Formula (2): **TOT.DF** $= N - 1 = 40 - 1 = 39$ DF.

Module III: Between-subject component

Step 3

(i) Formula (3): **B-S.SS**

$$= \frac{T_{S1}^2 + T_{S2}^2 + \ldots + T_{Sn}^2}{n_S} - CT$$

$$= \frac{37^2 + 45^2 + \ldots + 44^2}{8} - 1123{\cdot}6 = 5{\cdot}9$$

(ii) Formula (4): **B-S.DF** $= S - 1 = 5 - 1 = 4$ DF.

Module IV: Within-subject component

Step 4

(i) Formula (11): **W-S.SS** $=$ **TOT.SS – B-S.SS**

$= 246{\cdot}4 - 5{\cdot}9 = 240{\cdot}5$

(ii) Formula (12): **W-S.DF** $=$ **TOT.DF – B-S.DF**

$= 39 - 4 = 35$ DF.

Part 1:
Step 5

(i) Formula (5): **ME(C).SS**

$$= \frac{T_{C1}^2 + T_{C2}^2}{n_C} - CT$$

$$= \frac{98^2 + 114^2}{20} - 1123{\cdot}6 = 6{\cdot}4$$

(ii) Formula (6): **ME(C).DF** $= k - 1 = 2 - 1 = 1$ DF.

Step 6

(i) Formula (13*a*): **W-S.ERR.SS(*a*)**

$$= \frac{T_{C1S1}^2 + T_{C1S2}^2 + \ldots + T_{C2S5}^2}{n_{CS}} - (CT + ME(C).SS + B\text{-}S.SS)$$

$$= \frac{17^2 + 22^2 + \ldots + 27^2}{4} - (1123{\cdot}6 + 6{\cdot}4 + 5{\cdot}9)$$

$$= 1139 - 1135{\cdot}9 = 3{\cdot}1$$

(ii) Formula (13*b*): **W-S.ERR.DF(*a*)** $=$ **ME(C).DF × S – ME(A).DF**

$= (1 \times 5) - 1 = 4$ DF.

Table 9.13. – cont.

Part 2:
Step 7

(i) Formula (5): **ME(D).SS** $= \dfrac{T_{D1}^2 + T_{D2}^2 + T_{D3}^2 + T_{D4}^2}{n_D} - \mathrm{CT}$

$= \dfrac{24^2 + 40^2 + 64^2 + 84^2}{10} - 1123{\cdot}6$

$= 1332{\cdot}8 - 1123{\cdot}6 = 209{\cdot}2$

(ii) Formula (6): **ME(D).DF** $= k - 1 = 4 - 1 = 3$ DF.

Step 8

(i) Formula (13*a*): **W-S.ERR.SS(*b*)** $= \dfrac{T_{D1S1}^2 + T_{D1S2}^2 + \ldots + T_{D4S4}^2}{n_{DS}}$

$- (\mathrm{CT} + \mathrm{ME(D).SS} + \mathrm{B\text{-}S.SS})$

$= \dfrac{3^2 + 6^2 + \ldots + 18^2}{2}$

$- (1123{\cdot}6 + 209{\cdot}2 + 5{\cdot}9)$

$= 1345 - 1338{\cdot}7 = 6{\cdot}3$

(ii) Formula (13*b*): **W-S.ERR.DF(*b*)** $= \mathrm{ME(D).DF} \times S - \mathrm{ME(D).DF}$

$= (3 \times 5) - 3 = 12$ DF.

Part 3:
Step 9

(i) Formula (7): **INT(C × D).SS** $= \dfrac{T_{C1D1}^2 + T_{C1D2}^2 + \ldots + T_{C2D4}^2}{n_{CD}}$

$- (\mathrm{CT} + \mathrm{ME(C).SS} + \mathrm{ME(D).SS})$

$= \dfrac{13^2 + 19^2 + \ldots + 47^2}{5}$

$- (1123{\cdot}6 + 6{\cdot}4 + 209{\cdot}2)$

$= 1347{\cdot}2 - 1339{\cdot}2 = 8{\cdot}0$

(ii) Formula (8): **INT(C × D).DF** $= \mathrm{ME(C).DF} \times \mathrm{ME(D).DF}$

Step 10

(i) Formula (13*b*): **W-S.ERR.SS** = **W-S.SS – Rest.SS**

$= 240{\cdot}5 - (6{\cdot}4 + 3{\cdot}1 + 209{\cdot}2 + 6{\cdot}3 + 8) = 7{\cdot}5$

(ii) Formula (13*b*): **W-S.ERR.DF** = **W-S.DF – Rest.DF**

$= 35 - (1 + 4 + 3 + 12 + 3) = 12$ DF.

Table 9.13. – cont.

Step 11 Module V: Calculate the mean squares and F-ratios

(i) $\mathbf{MS(C)} = \dfrac{\mathbf{ME(C).SS}}{\mathbf{ME(C).DF}} = \dfrac{6{\cdot}4}{1} = 6{\cdot}4$

(ii) $\mathbf{MS(D)} = \dfrac{\mathbf{ME(D).SS}}{\mathbf{ME(D).DF}} = \dfrac{209{\cdot}2}{3} = 69{\cdot}73$

(iii) $\mathbf{MS(C \times D)} = \dfrac{\mathbf{INT(C \times D).SS}}{\mathbf{INT(C \times D).DF}} = \dfrac{8{\cdot}0}{3} = 2{\cdot}67$

(iv) $\mathbf{MS(ERR}\boldsymbol{a}\mathbf{)} = \dfrac{\mathbf{W\text{-}S.ERR.SS(}\boldsymbol{a}\mathbf{)}}{\mathbf{W\text{-}S.ERR.DF(}\boldsymbol{a}\mathbf{)}} = \dfrac{3{\cdot}1}{4} = 0{\cdot}77$

(v) $\mathbf{MS(ERR}\boldsymbol{b}\mathbf{)} = \dfrac{\mathbf{W\text{-}S.ERR.SS(}\boldsymbol{b}\mathbf{)}}{\mathbf{W\text{-}S.ERR.DF(}\boldsymbol{b}\mathbf{)}} = \dfrac{6{\cdot}3}{12} = 0{\cdot}52$

(vi) $\mathbf{MS(ERR}\boldsymbol{c}\mathbf{)} = \dfrac{\mathbf{W\text{-}S.ERR.SS(}\boldsymbol{c}\mathbf{)}}{\mathbf{W\text{-}S.ERR.DF(}\boldsymbol{c}\mathbf{)}} = \dfrac{7{\cdot}5}{12} = 0{\cdot}62$

(vii) $\mathbf{F\text{-}R(C)} = \dfrac{\mathbf{MS(C)}}{\mathbf{MS(ERR}\boldsymbol{a}\mathbf{)}} = \dfrac{6{\cdot}4}{0{\cdot}77} = 8{\cdot}31$

(viii) $\mathbf{F\text{-}R(D)} = \dfrac{\mathbf{MS(D)}}{\mathbf{MS(ERR}\boldsymbol{b}\mathbf{)}} = \dfrac{69{\cdot}73}{0{\cdot}52} = 134{\cdot}1$

(ix) $\mathbf{F\text{-}R(C \times D)} = \dfrac{\mathbf{MS(C \times D)}}{\mathbf{MS(ERR}\boldsymbol{c}\mathbf{)}} = \dfrac{2{\cdot}67}{0{\cdot}62} = 4{\cdot}3$

Step 12 Module VI: Interpretation

Comparisons and Trend Tests

One of the restrictions associated with main effects involving more than two levels is that *you cannot categorically state whether the mean of any one level is significantly different from the mean of any other level*. Equally, you might wish to make comparisons within an interaction and again the interaction term will not usually provide you with a direct comparison. Such precise questions can be answered but they do involve extra calculations. It may also be useful to test for trends in your data. Perhaps you may have predicted a linear or monotonic change across the levels of one of your variables. Alternatively you may have predicted a non-monotonic effect, like an inverted U.

Table 9.14. Our 2 between-subject and 2 within-subject split plot in modular form

Step	Module	Source	Formula
Step 1	Module I	(Data presentation)	(none)
Step 2	Module II	TOT.SS/DF	(1)/(2)
Step 3	Module III	B-S.SS/DF	(9)/(10)
Step 4		ME(A).SS/DF	(3)/(4)
Step 5		ME(B).SS/DF	(3)/(4)
Step 6		INT(A × B).SS/DF	(5)/(6)
Step 7		B.ERR.SS/DF	(7)/(8)
Step 8	Module IV	W-S.SS/DF	(11/(12)
Step 9	Part 1	ME(C).SS/DF	(3)/(4)
Step 10		INT(C × A).SS/DF	(5)/(6)
Step 11		INT(C × B).SS/DF	(5)/(6)
Step 12		INT(C × A × B).SS/DF	(5)/(6)
Step 13		Error(*a*).SS/DF	(13*a*)/(14*a*)
Step 14	Part 2	ME(D).SS/DF	(3)/(4)
Step 15		INT(D × A).SS/DF	(5)/(6)
Step 16		INT(D × B).SS/DF	(5)/(6)
Step 17		INT(D × A × B).SS/DF	(5)/(6)
Step 18		Error(*b*).SS/DF	(13*a*)/(14*a*)
Step 19	Part 3	INT(C × D).SS/DF	(5)/(6)
Step 20		INT(C × D × A).SS/DF	(5)/(6)
Step 21		INT(C × D × B).SS/DF	(5)/(6)
Step 22		INT(C × D × A × B).SS/DF	(5)/(6)
Step 23		Error(*c*).SS/DF	(13*b*)/(14*b*)
Step 24	Module V	MS/F-R	(15)/(16)
Step 25	Module VI	(interpretation)	(none)

Comparisons (both planned and unplanned) and trend tests rely on basically the same set of procedures which include the weighting of group *totals* by the use of coefficients. Such techniques are applied *after* the completion of the main ANOVA computation.

1 Planned comparisons

Planned comparisons are used when you wish to make comparisons between the means of either single levels or combinations of levels of variables. However, there are certain restrictions on their use. First, they are called *planned* comparisons because you must have decided to make such comparisons *before* the ANOVA was undertaken. Second, *the number of comparisons you make must not be greater than the number of degrees of freedom contained in the particular effect.* Therefore, if it is a main effect, with three degrees of freedom, you can undertake a maximum of three planned comparisons *each with one degree of freedom.*

The third restriction is more concerned with the actual computation. In a planned comparison the coefficients assigned to one half of the comparison (either one level or a combination of levels) are positive whilst those assigned to the other half are negative. The total weights for the two halves should also be equal and thus sum to zero. For example, if you wish to compare one level of a variable with the remaining three levels you would assign the coefficients as:

Levels of variable	$L1$	$L2$	$L3$	$L4$	
Coefficients	+3	−1	−1	−1	Sum = 0

Therefore for each comparison, or set of coefficients, the coefficients must *sum to zero* and this constraint will determine the actual weight of each coefficient.

In addition, if you wish to carry out more than one planned comparison you must ensure that all the sets of coefficients used are *mutually orthogonal*. This means that not only does each set of coefficients sum to zero, but the sum of the cross-products does as well. The reason why the coefficients need to be mutually orthogonal is that the comparisons must be independent of each other. That is, you must not use the same mean twice, in any one set of comparisons. To check this the coefficients assigned to each level are multiplied together and the product summed across all levels.

Table 9.15 illustrates an example where the coefficients are mutually orthogonal, and Table 9.16 an example of where they are not.

Having assigned the appropriate coefficients to the totals we wish to compare, we then apply the formula below to obtain the planned comparison sum of squares:

$$\text{P.C.SS} = \frac{(\Sigma ct)^2}{n\Sigma c^2}$$

where t = level or group total, c = weighted coefficient and n = the number of scores making up each total.

Table 9.15. An example of some mutually orthogonal coefficients

	L1	*L2*	*L3*	*L4*	
Comparison 1: Set 1 (*L1* vs *L2*)	+1	−1	0	0	Sum = 0
Comparison 2: Set 2 (*L3* vs *L4*)	0	0	+1	−1	Sum = 0
Cross-product	0	0	0	0	Sum = 0

The result should then be placed in your ANOVA summary table. You should note that the comparison's variance is part of the effect you are examining and therefore is *not* added to the total sum of squares. Where a full set of mutually orthogonal comparisons have been made, you should find that if all their sums of squares are added together they equal the sum of squares of the effect you are examining. This also applies to their degrees of freedom.

Table 9·16. An example of some coefficients which are not mutually orthogonal

	L1	*L2*	*L3*	*L4*	
Comparison 1: Set 1 (*L1* vs (*L2* + *L3*))	+2	−1	−1	0	Sum = 0
Comparison 2: Set 2 ((*L2* + *L3*) vs *L4*)	0	+1	+1	−2	Sum = 0
Cross-product	0	−1	−1	0	Sum = −2

Step-by-Step Procedure

Planned comparison

	Calculations	Usual symbol
Step 1	Tabulate group totals (one for each level).	$t_1, t_2 \ldots t_n$
Step 2	Assign weighted coefficients.	$c_1, c_2 \ldots c_n$
Step 3	Multiply each total by its coefficient.	$c_1t_1, c_2t_2 \ldots c_nt_n$
Step 4	Sum all the products from step 3.	Σct
Step 5	Square step 4.	$(\Sigma ct)^2$
Step 6	Work out the number of scores making up each total (all n should be equal).	n
Step 7	Square each coefficient.	$c_1^2, c_2^2 \ldots c_n^2$
Step 8	Sum all the coefficients from step 7.	Σc^2
Step 9	Multiply step 6 by step 8.	$n\Sigma c^2$
Step 10	Divide step 5 by step 9 to give PC.SS.	$\frac{(\Sigma ct)^2}{n\Sigma c^2}$
Step 11	Place result in summary table and calculate MS and F-R as usual.	

Worked Example

Planned comparison

Following a one-way simple factorial ANOVA, we wish to make a comparison between level 1 and level 2 of our independent variable 'noise'. Five scores had been collected in each level.

Steps 1 to 3

	$L1$	$L2$	$L3$	$L4$
t =	15	18	20	24
c =	+1	−1	0	0
ct =	+15	−18	0	0

Step 4 $\Sigma ct = -3$

Step 5 $(\Sigma ct)^2 = 9$

Step 6 $n = 5$

Step 7 $1^2 \quad 1^2 \quad 0^2 \quad 0^2$

Step 8 $\Sigma c^2 = 2$

Step 9 $n\Sigma c^2 = 5 \times 2 = 10$

Step 10 $\frac{9}{10} = 0{\cdot}9$

Step 11

Source	SS	DF	MS	F-R	P
ME(A)	25·35	3	8·45	7·04	$P < 0{\cdot}001$
PC	0·90	1	0·90	0·75	NS
Error	19·20	16	1·20		
Total	44·55	19			

2 Unplanned comparisons

There are occasions when you may want to make comparisons following an ANOVA, although you had not planned them. In this situation it is possible to use an unplanned comparison (Scheffes test). Unplanned comparisons are not subject to the restriction that the coefficients must be mutually orthogonal. However, you would *use the same formula and follow the same step-by-step procedures as planned comparisons*. The only difference is that before consulting the F-tables the F-ratio needs to be transformed to F' by dividing F by $(k - 1)$, i.e.

$$F' = \frac{F}{k-1}$$

where F' has $k - 1$ and $N - k$ degrees of freedom, k is the number of groups or levels and N is the total number of scores.

In addition, it is recommended that alpha (the significance level) is set at 0·1 with this test.

3 Trend tests

Trend tests are concerned with looking for trends shown by the shape or pattern of your means across levels within one variable. The number of possible trends increases with the number of levels or groups. It is not possible, for obvious reasons, to determine trends with only two levels or groups because two points cannot really indicate a trend. We therefore use trend tests for three or more groups. It is also not meaningful to test for trends with unequally spaced levels (e.g. 1, 10 and 50 rather than 2, 4 and 6). A linear trend is an increasing or decreasing straight line and a quadratic is more like a U or inverted-U shape. With four groups you can also test for a cubic trend (∿) and five groups a quartic trend (⩘). A trend test uses coefficients weighted to reflect the particular shape you wish to look for. Then, by using the same formula as planned comparisons you can determine whether you have a significant trend.

The appropriate weighted coefficients, according to the number of groups, for the different trends are outlined in Table 9.17. Note that the same restrictions apply to trend tests as did to planned comparisons, and that they likewise have *one degree of freedom*.

The procedures and formula for determining the F-values for

Table 9.17. Mutually orthogonal coefficients for ANOVA-based trend tests

Group size	Trend	Coefficients
3	Linear	−1 0 +1
	Quadratic	+1 −2 +1
4	Linear	−3 −1 +1 +3
	Quadratic	+1 −1 −1 +1
	Cubic	−1 +3 −3 +1
5	Linear	−2 −1 0 +1 +2
	Quadratic	−2 +1 +2 +1 −2
	Cubic	−1 +2 0 −2 +1
	Quartic	+1 −4 +6 −4 +1
6	Linear	−5 −3 −1 +1 +3 +5
	Quadratic	+5 −1 −4 −4 −1 +5
	Cubic	−5 +7 +4 −4 −7 +5
	Quartic	+1 −3 +2 +2 −3 +1

trend tests are the same as for the planned comparisons. However, you should note that one set of data may have more than one trend associated with it. For example, across the group totals below there is both a linear and a quadratic trend.

	L1	L2	L3		
Totals	19	36	29		
Linear	−1	0	+1		
Quadratic	−1	+2	−1		
Linear *ct*	−19	0	29	= 10	
Quadratic *ct*	−19	72	−29	= 24	$n = 5, k = 3$

so that

$$\text{Linear.SS} = \frac{(\Sigma ct)^2}{n\Sigma c^2} = \frac{(10)^2}{(5 \times 2)} = \frac{100}{10} = 10$$

$$\text{Quad.SS} = \frac{(\Sigma ct)^2}{n\Sigma c} = \frac{(24)^2}{(5 \times 6)} = \frac{576}{30} = 19{\cdot}2$$

Both sets of results should be placed in the ANOVA summary table (see Table 9.18).

Table 9.18. Summary table for the trend test example

Source	SS	DF	MS	F-R	*P*
ME(A)	29·2	2	14·6	10·73	0·001
Linear	10·00	1	10·00	7·35	0·05
Quadratic	19·2	1	19·2	14·12	0·001
Error	16·34	12	1·36		
Total	45·54	14			

Summary

This chapter outlined when and why ANOVA techniques are used, and discussed some of the advantages over more simple parametric tests. It then explored the rationale behind ANOVA and the basic computational formulae involved.

The final section explained how to carry out the ANOVA computation for three commonly used designs. This was achieved through the adoption of a modular system, rather than the simple step-by-step procedures used in other chapters. This reflects the complexity and number of necessary computational procedures involved in ANOVA. However, some summarized step-by-step procedures were provided, together with worked examples, in order to help those of you who may have found the modular approach difficult. Many of the computational modules are now commonly worked out by computers, thus emphasizing the importance of correct data presentation (Module I) and interpretation (Module VI). In addition to the three designs, further statistical techniques were provided for testing trends and making comparisons following the basic ANOVA computations.

Useful Computational ANOVA Formulae

(1) $\text{TOT.SS} = \Sigma X^2 - \dfrac{(\Sigma X)^2}{N}$

where

ΣX^2 = sum of the squares of each observation (X), i.e. square first then add.

$(\Sigma X)^2$ = the square of the sum of the observations, i.e. add first then square.

N = the total number of observations.

$\frac{(\Sigma X)^2}{N}$ = the correction term or CT

(2) TOT.DF $= N - 1$

(3) B-S.SS $= \frac{T_{s1}^2 + T_{s2}^2 + \ldots + T_{sn}^2}{n_s} - \text{CT}$

where $T_{s1}^2 + T_{s2}^2 + \ldots + T_{sn}^2$ = the sum of the squared subject totals and n_s = the number of scores which made up each subject total.

(4) B-S.DF $= S - 1$

where S = the number of subjects.

(5) ME(A).SS $= \frac{T_{A1}^2 + T_{A2}^2 + \ldots + T_{An}^2}{n_A} - \text{CT}$

where $T_{A1}^2 + T_{A2}^2 + \ldots + T_{An}^2$ = the sum of the squares of the totals for each level in the main effect, and n_A = the number of observations which made up *each* of the above totals.

(6) ME(A).DF $= k - 1$

where k = the number of levels in variable A.

(7) INT(A × N).SS $= \frac{T_{A1N1}^2 + T_{A1N2}^2 + \ldots T_{AnNn}^2}{n_{AN}} - (\text{CT} + \text{ME(A} \ldots \text{N).SS} + \text{INT(A} \times \ldots \text{N).SS})$

where:

$T_{A1N1}^2 + T_{A1N2}^2 + \ldots + T_{AnNn}^2$ = the sum of the squared interaction totals for all combinations of A to N, at all levels.

ME(A ... N).SS = all the main effects subsumed within the interaction. For example, a two-way interaction (A × B) has two main effects (A and B). A three-way interaction (A × B × C) has three main effects (A, B, C) subsumed and so on.

INT(A × ... N).SS = all the interaction effects subsumed within the interaction being calculated. For example, a two-way interaction has no other interactions subsumed. A three-way interaction (A × B × C) has three two-way interactions subsumed (A × B; A × C; B × C). A four-way interaction (A × B × C × D) has six two-way interactions ($A \times B$, $A \times C$, $A \times D$, $B \times C$, $B \times D$ and $C \times D$) and four three-way interactions ($A \times B \times C$, $A \times B \times D$, $A \times C \times D$ and $B \times C \times D$).

n_{AN} = the number of observations which make up each of the interaction totals, and CT = correction term.

(8) INT(A × N).DF = ME(A).DF × . . . ME(N).DF

(9) B.ERR.SS = B-S.TOT − (ME(A . . . N).SS + INT(A × . . . N).SS)

where:

B-S.TOT = the between-subject total
ME(A . . . N).SS = all the between-subject main effects
INT(A × . . . N).SS = all the between-subject interactions

(10) B.ERR.DF = B-S.DF − (ME(A . . . N).DF + INT(A × . . . N).DF)

(11) W-S.SS = TOT.SS − B-S.SS

where TOT.SS = the total sum of squares
B-S.SS = the between-subject sums of squares

(12) W-S.DF = TOT.DF − B-S.DF

(13a) W-S.ERR.SS(1) = $\dfrac{\text{(ME(A)/INT(A × N).SS × S)}}{n\text{ME(A)/INT(A × N)}}$ − (CT + ME(A . . . N)/INT(A × . . . N).SS + B-S.SS)

(14a) W-S.ERR.DF(1) = ME(A)/INT(A × N).DF × S − ME(A . . . N)/INT(A × . . . N).DF

(13b) W-S.ERR.SS(2) = W-S.SS − all other within-subject SS

(14b) W-S.ERR.DF(2) = W-S.DF − all other within-subject DFs

(15) ME(A).MS = $\dfrac{\text{ME(A).SS}}{\text{ME(A).DF}}$ and

INT(A × N).MS = $\dfrac{\text{INT(A × N).SS}}{\text{INT(A × N).DF}}$

(16) F-R(A) = $\dfrac{\text{ME(A).MS}}{\text{ERROR.MS}}$ or

F-R(A × N) = $\dfrac{\text{INT(A × N).MS}}{\text{ERROR.MS}}$

10 Non-Parametric Analysis of Variance by Ranks

There may be occasions when it is not advisable to use a parametric test, because you have breached one or more of the fundamental assumptions underpinning the use of such tests. In these situations there may be an appropriate non-parametric test that you can use. However, we do not wish to give the impression that a non-parametric test should only be used when the equivalent parametric test cannot be. Non-parametric tests have advantages in their own right. These advantages, together with some disadvantages, have already been discussed in Chapters 3 and 5.

Some non-parametric tests are used on ordinal data, for comparing two levels of a variable or two groups. A number of these sorts of tests will already be familiar to you. For example, the Wilcoxon test can be used for comparing two groups in a within-subject or matched-pair design and the Mann–Whitney test can be used to compare two groups in a between-subject design. Other non-parametric tests can deal with more than two levels or groups incorporated in a within-subject or between-subject design.

In this chapter, we will introduce you to two of these more complex non-parametric tests. They are essentially analogous to the parametric one-way ANOVAs. However, despite their name, they do not actually analyse variance. The name *non-parametric ANOVA by ranks* is used to signify certain design similarities with ANOVA. In addition, we will examine the non-parametric equivalents of linear trend tests and one test of interactions.

The Kruskal–Wallis Test for Between-subject (One-way) Designs

The Kruskal–Wallis one-way ANOVA is used for comparing three or more levels of a variable in a *between-subject design*, and can be used in similar situations to those in which its parametric equivalent (one-way simple factorial) is used. It is worth noting that compared to the one-way simple factorial ANOVA, the Kruskal–Wallis test has a power efficiency of 95·5 per cent.

The logic behind the Kruskal–Wallis test is quite simple, and forms the basis of most non-parametric tests using ranked data.* Imagine that we have three groups containing $(n_1 + n_2 + n_3 = N)$ and for simplicity's sake the groups are all of equal size. If we rank our scores, regardless of group membership (i.e. 1 to N), then under the null hypothesis we would expect the sum of the ranks for each group to be approximately equal.

If the scores in group one are low, in group two middling and in group three high, then evidently in this situation the sum of the ranks for each group would not be equal. If the sums of the ranks for the three samples are sufficiently different to give a significant effect, we can assume that the samples were drawn from different populations. Consequently we can reject the null hypothesis in favour of our alternative hypothesis. This does assume, however, that the samples were drawn randomly from their respective populations, and that these populations have the same distribution characteristics (e.g. means and variances). Therefore this and similar non-parametric tests using ranks do make some quite stringent assumptions.

As a first step in the Kruskal–Wallis computation, it is advisable to place your data in a factorial matrix similar to the one described for the equivalent ANOVA in Chapter 9. In order then to compute the crucial statistic H, your data need to be measured on an ordinal scale. If your data are not already in this form then you will need to rank all the scores, regardless of group membership. As already stated, the sum of the ranks for each group should be approximately equal under the null hypothesis. The size of H indicates by how much the true sum differs from our expectation under the null hypothesis.

We compute H by the use of the following formula:

*In fact the Kruskal–Wallis test can be thought of as a generalized version of Mann–Whitney's test where

$$U = n_1 n_2 + \frac{n_1(n_1 + 1)}{2} - R_1 \quad \text{or} \quad n_1 n_2 + \frac{n_2(n_2 + 1)}{2} - R_2$$

where R_1 = sum of ranks of the group with n_1 sample size, and R_2 = sum of ranks of the group with n_2 sample size.

$$H = \frac{12}{N(N+1)} \sum_{i=1}^{k} \left(\frac{R_i^2}{n_i} \right) - 3(N+1)$$

where N = total sample size
n = number of observations in a given group
R = sum of the ranks in a given group
k = number of groups.

If there are ties, then you should adopt the usual procedure, as shown in Chapter 6 (p. 104), of assigning to the tied observations the average of the ranks they would otherwise occupy. Then divide H by the following:

$$1 - \frac{\Sigma T}{N^3 - N}$$

where $T = t^3 - t$ and t is the number of tied observations in a given rank.

Step-by-Step Computational Procedures

Kruskal–Wallis test

Step	Calculations	Usual symbols
Step 1	Rank all scores regardless of group membership (1 to N); assign to tied observations the average of the ranks they would otherwise occupy.	
Step 2	Note sample size.	N

Worked Example

Kruskal–Wallis test

You may wish to test whether social class has a significant effect on people's feelings towards accepting unemployment benefit. To do this, you could select three social class groups (e.g. group I: upper middle class; group II: lower middle class; and group III: working class). You could then ask a number of persons from each group to rate their feelings on a 20-point scale. A high score would indicate a favourable response to the acceptance of benefits. The sort of raw data you may obtain is in Table 10.1 below. Notice that different sample sizes can be used. In the present example, the numbers of ratings from each social class are four, four and five respectively.

Table 10.1. Favourability rating towards the question of accepting unemployment benefit

Social class		
I	II	III
16	8	4
17	10	9
20	10	5
16	11	7
		10

Step 1

	Social class		
	I	II	III
	10·5	4	1
	12	7	5
	13	7	2
	10·5	9	3
			7
$R =$	46	27	18

Step 2 $N = 13$

Step	**Calculations**	**Usual symbols**
Step 3	Note the number of observations in each group.	$n_1, n_2 \ldots n_i$
Step 4	Find the sum of the ranks for each group.	$R_1, R_2 \ldots R_i$
Step 5	Square each sum of ranks.	$R_1^2, R_2^2 \ldots R_i^2$
Step 6	Add 1 to step 1 and multiply by step 1.	$N(N+1)$
Step 7	Divide 12 by step 6.	$\frac{12}{N(N+1)}$
Step 8	Divide the square of each sum of the ranks for each group by the number of observations in that group; do this by dividing each sum in step 5 by the associated number of observations in step 3.	$\frac{R_1^2}{n_1}, \frac{R_2^2}{n_2} \ldots \frac{R_i^2}{n_i}$
Step 9	Sum the result of step 8.	$\sum \frac{R_i^2}{n_i}$
Step 10	Add 1 to step 1 and multiply by 3.	$3(N+1)$
Step 11	Multiply step 7 by step 9 and subtract step 10.	$H = \frac{12}{N(N+1)} \sum \frac{R_i^2}{n_i} - 3(N+1)$
Step 12a	*If your data contain ties* then count the number of ties for a given rank (*f*).	$t_1, t_2 \ldots t_i$

Step 3 $n_1 = 4, n_2 = 4, n_3 = 5$

Step 4 $R_1 = 46, R_2 = 27, R_3 = 18$

Step 5 $R_1^2 = 2116, R_2^2 = 729, R_3^2 = 324$

Step 6 $N(N+1) = 13(13+1) = 182$

Step 7 $\frac{12}{N(N+1)} = \frac{12}{182} = 0{\cdot}0659$

Step 8 $\frac{46^2}{4}, \frac{27^2}{4}, \frac{18^2}{5}$

Step 9 $\sum \frac{R_i^2}{n_i} = 776{\cdot}05$

Step 10 $3(N+1) = 42$

Step 11 $H = (0{\cdot}0659)(776{\cdot}05) - 42 = 9{\cdot}14$

Step 12a There are ties: Rank 7 = 3 ties
Rank 10·5 = 2 ties

Step	**Calculations**	**Usual symbols**
Step 12b	Calculate the cubes of step 12*a*.	$t_1^3, t_2^3 \ldots t_i^3$
Step 12c	Subtract step 12*a* from step 12*b* for associated values and sum these results.	$\Sigma(t^3 - t)$
Step 12d	Find the cube of step 1 and subtract step 1 from this result.	$N^3 - N$
Step 12e	Divide step 12*c* by step 12*d*.	$\frac{\Sigma(t^3 - t)}{N^3 - N}$
Step 12f	Subtract the result of step 12*e* from 1.	$1 - \frac{\Sigma(t^3 - t)}{N^3 - N}$
Step 12g	Subtract step 12*f* from step 11 to correct for ties.	*H* corrected for ties
Step 13	Assess the significance of H when $k = 3$ and $n = 1, 2, 3$ or 4 by using Table V in the Appendix. In other cases use Table VI, with $k - 1$ degrees of freedom.	
Step 14	Inference. If the value of H is equal to or greater than the critical value at the given significance level then the null hypothesis can be rejected in favour of your alternative hypothesis (i.e. you have found a significant effect).	

Step 12b $t^3 = 27,\ t^3 = 8$

Step 12c $\Sigma(t^3 - t) = (27 - 3) + (8 - 2) = 30$

Step 12d $(N^3 - N) = 2184$

Step 12e $\dfrac{\Sigma(t^3 - t)}{N^3 - N} = 0{\cdot}014$

Step 12f $1 - \dfrac{\Sigma(t^3 - t)}{N^3 - N} = 1 - 0{\cdot}014 = 0{\cdot}986$

Step 12g H corrected for ties $= \dfrac{9{\cdot}14}{0{\cdot}986} = 9{\cdot}267$

Step 13 In our example $k = 3$ and ns are 4, 4 and 5, so we can then use Table V to look up the significance of our H value. From Table V we can see that the probability associated with our H value is $P < 0{\cdot}009$. The effect is therefore very significant.

Friedman's Test for Within-subject (One-way) Designs

The Friedman test is based on a similar logic to the Kruskal–Wallis test, but it is used in correlated or *within-subject* one-way designs. Recall that in a within-subject design the same subject, or *matched* group of subjects, undergoes all the different treatments or levels of one independent variable. The Friedman test differs from its parametric equivalent (repeated measures ANOVA design) in that it is applied to ordinal-level data, but is similar in that it is particularly sensitive to changes in central tendency.

A basic data matrix should be used to organize your data into treatment levels, which are represented by the columns. Subjects (or matched groups) are represented by the rows. If we have four treatments ($k = 4$) and eight subjects ($n = 8$) then we would have a 4×8 matrix. We then rank each row *separately* (i.e. in this case from

---

Step-by-Step Procedure

Friedman's test

Steps	Calculations	Usual symbols
Step 1	Place the data in a two-way table with k columns and n rows.	

1 to 4). If the null hypothesis was true we would expect our ranks 1 to 4 to appear with about equal frequency in all columns. Our column totals should therefore be about equal. If, however, the totals are shown to be statistically different then we can reject the null hypothesis. The power efficiency of Friedman's test varies from 64 per cent for $k = 2$ to a maximum of 95·5 per cent for infinitely large numbers of ks.

The Friedman test determines whether there are significant effects through the application of the following formula:

$$\chi_r^2 = \frac{12}{nk(k+1)} \sum_{i=1}^{k} (R_i^2) - 3n(k+1)$$

where n = number of rows, k = number of columns and R_j = sum of the ranks of jth column.

Worked Example

Friedman's test

You may be interested in finding out if there is a significant effect from using three different programmes for the teaching of mathematics to children. To test this you could select five groups of three children, each group containing children matched for intellectual ability. The dependent variable would be the difference between a pre- and post-score on a test of mathematical ability. It could be assumed that the larger the increase in score, the greater the effectiveness of the particular teaching programme. The data obtained could be similar to that set out below:

Step 1

		Teaching programmes		
		I	II	III
Matched groups	1	0	8	2
	2	3	5	2
	3	5	6	4
	4	0	8	3
	5	6	4	5

Steps	Calculations	Usual symbols
Step 2	Rank the scores in each row (1 to n).	
Step 3	Sum the ranks of each column and square each sum.	$R_1, R_2 \ldots R_i; R_1^2, R_2^2 \ldots R_i^2$
Step 4	Add 1 to the number of columns and multiply this sum by the number of columns and then by the number of rows.	$nk(k+1)$
Step 5	Divide 12 by step 4.	$\frac{12}{nk(k+1)}$
Step 6	Add together the squared sum of ranks for each column.	ΣR_i^2
Step 7	Multiply step 5 by step 6.	$\frac{12}{nk(k+1)} \Sigma(R_i^2)$

Step 2

	Teaching programmes		
	I	II	III
1	1	3	2
2	2	3	1
3	2	3	1
4	1	3	2
5	3	1	2
R	9	13	8

Step 3 9, 13, 18; 81, 169, 324

Step 4 $(3)(5)(3+1) = 60$

Step 5 0·2

Step 6 $81 + 169 + 324 = 574$

Step 7 $0{\cdot}2 \times 574 = 114{\cdot}8$

Steps	Calculations	Usual symbols
Step 8	Add 1 to the number of columns and multiply this by the number of rows and then by 3.	$3n(k+1)$
Step 9	Subtract step 8 from step 7 to give χ_r^2	$\chi_r^2 = \frac{12}{nk(k+1)}\Sigma(R_i^2) - 3n(k+1)$
Step 10	Assess the significance of χ_r^2 when $k = 3$ and n is less than 9 or $k = 4$ and n is less than 4 by using Table VII in the Appendix; in other cases use Table VI in the Appendix (with $k - 1$ DF).	
Step 11	Decide whether to accept or reject the null hypothesis. If the value of χ_r^2 is equal to or greater than the critical value at a given significance level then the null hypothesis can be rejected in favour of the alternative hypothesis (i.e. you have found a significant effect).	

Non-parametric Trend Tests

In Chapter 9 we discussed the trend tests that could be applied to parametric ANOVA data. In this section, we will describe two non-parametric trend tests. The first of these is a test for linear trends across groups in a between-subject design, and the second is a similar technique for a within-subject design.

1 Between-subject design (Jonckheere's test)

This trend test can be used in association with the Kruskal–Wallis test discussed earlier. It involves the statistic *S*, as employed by Kendall's tau.* It can be used to test for linear or monotonic trend,

* Kendall's tau is not itself discussed in this book since it is equivalent to Spearman's rho, which was described in Chapter 6.

Step 8 $(3)(5)(3 + 1) = 60$

Step 9 $114{\cdot}8 - 60 = 54{\cdot}8$

Step 10 $P < 0{\cdot}001$

Step 11 Reject the null hypothesis, the group effect is significant.

---

across three or more independent samples, each sample having been given a different level of a particular treatment.

The data below represent possible scores from each of the three samples. The scores could again be the number of errors made in typing a letter. Each group or sample may have performed the task under different levels of noise.

Sample I	Sample II	Sample III
2	5	4
8	7	9
6	11	13
10	12	15
		$N = 12$

The test does not require ordinal data, but the samples should be arranged so that the sample for which you predict the lowest scores appears as the first column through to the sample with the predicted highest scores as the last column. The order of the samples is thus arranged to reflect the trend you are testing for.

Then, you need to calculate S where

$$(4) \quad S = N^+ - N^-$$

You determine N^+ by taking the first score in sample I (2) and counting the number of scores to the right of it which are greater than that score (i.e. in the other two samples). All the scores in samples II and III are greater than 2, so for the first score we have a figure of 8. You repeat the procedure for all the remaining scores in sample I, yielding figures of 5, 6 and 4 respectively.

Repeat the procedure with the second sample by counting the number of scores in sample III greater than each score in sample II (3, 3, 2 and 2). Then sum all the values obtained to find N^+:

$$N^+ = (8 + 5 + 6 + 4) + (3 + 3 + 2 + 2) = 33$$

To compute N^- you repeat the procedure but this time count all those scores which are *smaller* than a particular score.

$$N^- = (0 + 3 + 2 + 4) + (1 + 1 + 2 + 2) = 15$$

You can now compute a value for S:

$$S = 33 - 15 = 18$$

Next, you need to compute the normal deviate z:

$$(5) \quad z = \frac{S_c}{\sqrt{\mathrm{Var}S}}$$

where

$$S_c = S - 1$$

and then

$$(6) \quad \mathrm{Var}S = \frac{2(N^3 - \Sigma t^3) + 3(N^2 - t^2)}{18}$$

where t = number of scores in one sample.

So that from this example:

$$\text{Var}S = \frac{2(1728 - 192) + 3(144 - 48)}{18} = 186{\cdot}67$$

$$S_c = 18 - 1 = 17$$

Therefore

$$z = \frac{17}{\sqrt{186{\cdot}67}} = 1{\cdot}24$$

By using Table I in the Appendix we can see that the probability associated with a $z = 1{\cdot}24$ is 0·1075, which (at $\alpha = 0{\cdot}05$) is not significant.

If you have ties, then you should assign the numeral in the direction of your predicted trend. For example, if the first value of sample II had been 2, thus tying with the first value of sample I, you would treat it as if it was larger. You will then need to assign +1 and −1 for determining N^+ and N^- respectively. However, the test will be more conservative under these conditions.

Summary of the procedure:

Step 1 Arrange treatments or samples in the order which reflects the predicted trend.

Step 2 Calculate S by applying formula (4), correcting for ties.

Step 3 Calculate the standard error of S by formula (6).

Step 4 Calculate the normal deviate z by using formula (5).

Step 5 Check the significance by looking up the probability associated with that value of z in Table I (see Appendix).

Step 6 On the bases of step 5 decide whether to reject or accept the null hypothesis.

Within-subject design (Page's L)

Page's L is a linear trend test for within-subject (or correlated) designs, where the data are measured on an ordinal scale. It is usually used in associated with Friedman's related ANOVA.

Table 10.2 is an example of the possible scores from four sets of matched subjects, each incorporating one subject from each treatment group. Suppose you predict that there will be a linear increase from group I to III.

Table 10.2. Hypothetical scores of four sets of matched subjects across treatments

	Treatments I	II	III
1	2(1)	5(3)	4(2)
2	8(1·5)	8(1·5)	9(3)
3	6(1)	11(2)	13(3)
4	4(1)	12(2)	15(3)
R	4·5	8·5	11
y	1	2	3
y_R	4·5	17	33

(Ranks 1 to k in parentheses)

First, just as with Friedman's test, you start by placing the data into a two-way table like the one above and then you rank the scores across rows (i.e. from 1 to k). Second, you need to sum the ranks for each column (R). Third, you assign a numeral (y) to each column from 1 to k for an increasing trend, and k to 1 for a decreasing trend. Fourth, you need to calculate L by using the formula below:

$$(6) \quad L = \Sigma y_R$$

In our example:

$$L = (4{\cdot}5 + 17 + 33) = 54{\cdot}5$$

Finally, you can look up the critical values of L in the Appendix (Table VIII) for $k = 3$, $n = 2$ to 6, and $k = 4$, $n = 2$ to 5. In other cases, L can be transformed into z by using the formula below and the significance of z checked in Table I in the Appendix.

$$(7) \quad z = \frac{12L - 3n(k+1)^2}{k\sqrt{n(k-1)(k+1)}}$$

In our example, with $k = 3$ and $n = 4$, we would need an L value *larger* than 54 in order for it to be significant. This would be at the $P < 0{\cdot}05$ level with a one-tail test. Our value of 54·5 is thus just significant.

Summary of the procedure

Step 1 Place the data in a two-way table with k columns and N rows.

Step 2 Rank the observations across rows (1 to k).

Step 3 Sum the ranks for each column (R).

Step 4 Assign a numeral (y) from 1 to k or k to 1 for testing ascending and descending trends respectively.

Step 5 Compute L (formula (6)).

Step 6 Check for significance in Table VIII, or convert L into z (formula (7)) and use Table I.

Step 7 Decide to reject or accept the null hypothesis on the basis of step 6.

Non-parametric Analysis of Interactions

In Chapter 5 we pointed out that there are no readily available non-parametric techniques for analysing interactional data. However, there are some situations where interactional data can be analysed by using a combination of standard and familiar non-parametric techniques.

Such an analysis is based on the polynominal coefficients we discussed in Chapter 9. Recall that a strong interaction is epitomized by opposing (thus crossing) linear trends (e.g. see Figure 9.2). These trends can be represented by our coefficients. For example, if you had three levels in one variable (A1, A2 and A3) and two levels in a second variable (B1 and B2), then you could test for an interaction by a *linear by linear comparison* by using the following coefficients:

	A1	A2	A3
B1	+1	0	−1
B2	−1	0	+1

By applying the formula for testing trends which we presented in Chapter 9 (pp. 190–93) we could test whether the linear trend from A1 to A3 is significantly different for our two levels of B. However, usually such a procedure requires a parametric ANOVA to be undertaken first of all, although a Mann–Whitney's *U* test can be substituted *under certain conditions*.

For example, imagine that we have a split-plot (one between-subject and one within-subject) design consisting of data (not ranked) from eight subjects:

		A1	A2	A3
B1	1	10	6	1
	2	8	8	4
	3	7	3	2
	4	6	5	4
B2	5	8	7	6
	6	9	9	8
	7	4	2	4
	8	9	5	2

First, note that our coefficient table can be seen as simply A1B1 – A3B1 = A3B2 – A1B2. Therefore, if we do these calculations for each subject in B1 and B2 then we end up with two columns:

B1 A1 – A3	B2 A3 – A1
+9(8)	−2(2)
+4(6)	−1(3)
+5(7)	0(4)
+2(5)	−7(1)

(Ranks in parentheses)

We could then rank these scores and compare the two columns using Mann–Whitney's *U* as normal (see Robson, pp. 120–24, for detail of computing this test). If there is a significant difference between the groups then we can conclude that performance across levels of A varies with the level of B, i.e. that there is a significant A × B interaction.

The technique, in this form, works *only* for split-plot designs. Also when the number of variables or levels is greater than three, choosing and applying the correct coefficients is less simple and care should be taken. Finally, note that the between-subject main effects could have been independently assessed by a Kruskal–Wallis test (or a Mann–Whitney *U*) and the within-subject main effects by Friedman's test.

Summary

In this chapter we presented the rationale and worked examples for the non-parametric equivalents of the one-way simple factorial and repeated measures ANOVA designs. These were the Kruskal–Wallis one-way ANOVA and the Friedman's test respectively. We also detailed two non-parametric linear trend tests for use in between-subject and within-subject designs.

Finally we explored, in a limited way, a technique for testing interactions using non-parametric statistics.

11 Multivariate Techniques

This chapter describes several techniques that allow you to investigate the interrelationships of scores on *more than two* variables. For this reason the techniques are termed methods of *multivariate analysis*. Techniques to analyse the interrelationship between two variables, such as those considered in Chapter 6, are methods of *bivariate analysis*.

Partial Correlation

Partial correlation allows you to measure the relationship between two variables while eliminating or removing the effect of a third variable. Imagine that you have administered tests of statistical and creative ability to fifty children between the ages of eleven to sixteen and you obtain a correlation of 0·75 between the two tests. Are statistical and creative ability two sides of the same coin? Since statistical and creative ability may both increase with age you may reason that the correlation between statistical (V_1) and creative ability (V_2) may be to some extent dependent on the third variable, age (V_3).

Partial correlation works by using linear regression to see how well scores on V_1 can be predicted from V_3, and similarly for predictions of V_2 from V_3. The correlation between the error of estimate in predicting V_1 from V_3 and V_2 from V_3 is the *partial correlation coefficient*. It is the part of the correlation between V_1 and V_2 which remains when the effect of V_3 is eliminated, and it is numerically represented as $r_{12\cdot3}$. The formula for calculating partial correlation is:

$$r_{12\cdot3} = \frac{r_{12} - r_{13}r_{23}}{\sqrt{(1 - r_{13}{}^2)(1 - r_{23}{}^2)}}$$

For example, consider the correlation matrix presented in Table 11.1.

Table 11.1. A correlation matrix

		V_1	V_2	V_3
Statistical ability	(V_1)	1·00	0·75	0·60
Creative ability	(V_2)	0·75	1·00	0·55
Age	(V_3)	0·60	0·55	1·00

From this matrix

$$r_{12\cdot3} = \frac{0{\cdot}75 - (0{\cdot}60)(0{\cdot}55)}{\sqrt{[1 - (0{\cdot}6)^2][1 - (0{\cdot}55)^2]}}$$

$$= \frac{0{\cdot}75 - 0{\cdot}33}{\sqrt{(1 - 0{\cdot}360)(1 - 0{\cdot}303)}}$$

$$= \frac{0{\cdot}42}{\sqrt{(0{\cdot}640)(0{\cdot}697)}}$$

$$= \frac{0{\cdot}42}{\sqrt{0{\cdot}446}}$$

$$= \frac{0{\cdot}42}{0{\cdot}67} = \underline{\underline{0{\cdot}63}}$$

Is the partial correlation coefficient significantly different from zero? A t-test is used to test for significance, where

$$t = \frac{r_{12\cdot3}}{\sqrt{[1 - (r_{12\cdot3})^2]/(N-3)}}$$

with $N - 3$ degrees of freedom. In our example

$$t = \frac{0{\cdot}63}{\sqrt{(1 - 0{\cdot}397)/47}}$$

$$= \frac{0{\cdot}63}{\sqrt{0{\cdot}013}} = \underline{\underline{5{\cdot}562}}$$

Using the t tables and entering them with 47 degrees of freedom this result is significant at the 0·001 level.

Of course, another way to approach the problem would have been to match your sample of fifty subjects in terms of age! It is also possible to use partial correlation to remove the effects of more than one variable but because of difficulties of interpretation such coefficients are seldom calculated.

Multiple Regression and Correlation

Multiple regression allows you to study the relationship between *more than one* 'predictor' variable and a single criterion that is to be 'predicted'. Sometimes the scores on the criterion may be known beforehand and your interest may be in studying the relationship of scores on this variable to scores on other variables that are also known.

Multiple linear regression is an extension of simple linear regression which we discussed in Chapter 7. But instead of a single predictor variable, X, and a single criterion variable, Y, we have multiple predictor variables, e.g. X_1, X_2, X_3 . . . X_K. One use of multiple linear regression is to help us decide which of many potentially useful predictors are, in fact, useful to us in making a prediction of Y.

In multiple linear regression the fundamental linear model is:

$$Y = a + b_1X_1 + b_2X_2 + b_3X_3 \ldots + b_KX_K$$

The symbols have the same meaning as those in simple linear regression except that there are K predictor variables coupled with K regression coefficients or weights – the b values.

How should the scores in each of the X_K variables be combined? For instance, what weightings (b values) should be given to IQ score, first-year undergraduate mark, measurement of achievement motivation and a composite of 'A'-level grades in predicting final degree marks? In terms of simple linear regression one could choose as the *single* predictor variable the variable which correlates most highly with final degree mark. However, by computing a composite of scores on *all* the predictor variables one might be able to calculate a better estimate of final degree grade. Obviously, there is an unlimited number of ways that scores on the predictor variables could be combined. The simplest composite would be to sum the scores on the X_K variables. In most instances, however, a better composite could be obtained by giving some scores more weight than others. The problem is to find a system of weightings which will produce a composite that will correlate more highly with final degree grade than will any other composite. An approximate solution to this problem could be found by trial and error but the problem is solved best by the method of *multiple correlation* which weights the X_K scores to *maximize* the correlation between Y and their weighted sum. In essence, the a and b_K coefficients are selected by the least squares criterion in a similar way to the method involved in simple linear regression. The multiple correlation coefficient (R) is a measure of prediction accuracy for a particular sample and it can be interpreted in much the same way as an ordinary correlation coefficient, except that values of R range from 0 to 1 rather than from −1 to 1.

Analogous to our previous interpretation of r^2 in Chapter 7, R^2 is the proportion of the variance in the criterion variable, Y, accounted for by the predictor variables, X_K.

An F-ratio can be used to test whether an obtained multiple correlation coefficient is significantly different from zero. Where

$$F = \frac{R^2(N - g - 1)}{(1 - R^2)g}$$

and R = the multiple correlation coefficient, N = the number of observations and g = the number of predictors in the prediction equation. Enter the F tables on p. 256 with $df_1 = g$ and $df_2 = N - g - 1$.

For example, if an obtained R was 0·645 and this was based on a sample of fifty people's scores on five predictor variables,

$$F = \frac{0{\cdot}416(50 - 5 - 1)}{(1 - 0{\cdot}416)5} = \frac{18{\cdot}304}{2{\cdot}92}$$

$$= \underline{\underline{6{\cdot}27}}\ (P < {\cdot}01, \text{two-tailed})$$

Can the b values in the multiple regression equation be taken as weights that indicate the relative importance of the predictor variables in the prediction equation? No, although this would seem an obvious intuitive deduction: *the b values are not stable and there is no absolute way to interpret them*. If the correlations between the predictors were all zero and the b values were computed for standardized predictor scores, to control for scaling effects, we *could* talk of the relative importance of predictor variables. But in real-life many or most of the predictor variables that are correlated with the criterion variable are also correlated amongst themselves or, in other words, are to some extent redundant. Indeed, order of entry of the predictors into the prediction equation can be very important when these variables are to some extent correlated. The relative amount of variance in the criterion variable for which each predictor variable accounts will vary with order of entry.

This fact is particularly important when we wish to define the most parsimonious prediction equation such that only those predictors contributing greatly to prediction are included in it. Additional predictors contributing small and diminishing amounts to prediction should be eliminated from the prediction equation. The procedure for accomplishing this is called *stepwise multiple regression*. Here the best predictor is paired with every other predictor in turn and a multiple correlation is computed for each pair of predictors. When the best pair has been selected this pair is then combined with each of the remaining predictors and further multiple correlations are computed. The process continues until the predictors are ordered in terms of their relative contributions to prediction of the criterion variable. Changes in the multiple correlation coefficient are examined after each predictor is added to the prediction equation and tests of significance can be used to evaluate whether the multiple correlation has changed significantly. It is also possible to select, on theoretical grounds, the important predictors, and then selectively remove these and observe changes in the multiple correlation.

Often in psychological and educational research the criterion variable is not measured on an interval-level scale. It is often nominal, as with, for example, studies of those students who either pass or fail their final degree examination, or of those children who either pass or fail the eleven-plus. These types of prediction involve standard multiple linear regression analysis *but* with *point-biserial correlations* computed between the interval-level predictor variables and the nominal-level criterion variable. Such an analysis is often termed *discriminant function analysis* or, simply, *discriminant analysis*. The point-biserial correlation is simply the familiar product-moment correlation but with weights of, say, 1 or 0 assigned to one or other of the nominal categories. The coefficient is not affected by which particular weight you assign to which particular category *as long as you are consistent in your assignments*.

Once you have computed a discriminant analysis you will find that the *predicted* scores on the nominal criterion variable are, in fact, continuous and do not resemble the nominal data you used to construct the prediction equation. It follows that you will have to decide on a cut-off score. Your decision should follow from our discussion of these scores in Chapter 6.

Multiple regression: illustrative examples

One of the major data bases used for experimentation with multiple linear regression has been that collected by Meehl (1969). The judgemental problem used was that of differentiating psychotic from neurotic patients using their MMPI personality profiles.

Each patient upon being admitted to hospital had taken the MMPI. Expert clinical psychologists believe (or at least used to believe) that they can differentiate between psychotics and neurotics on the basis of a *profile* of the eleven scores. Meehl noted that:

> because the differences between psychotic and neurotic profile are considered in MMPI lore to be highly configural in character, an atomistic treatment by combining scales linearly should be theoretically a very poor substitute for the configural approach.

Initially researchers tried to 'capture' or 'model' expert judges by a multiple linear regression equation. These judgemental representations are constructed in the following fashion. The clinician is asked to make his diagnostic or prognostic judgement from a previously quantified set of cues for each of a large number of patients. These judgements are then used as the criterion variable in a standard linear regression analysis. The predictor variables in this analysis are the values of the cues. The results of such an analysis are a set of regression weights, one for each cue, and these sets of regression weights are referred to as the judge's 'model' or his 'policy'.

How do these models make out as predictors themselves? That is, if the regression weights (generated from an analysis of one clinical judge) were used to obtain a 'predicted score' for each patient, would these scores be more valid, or less valid, that the original clinical judgements from which the regression weights were derived? To the extent that the model fails to capture valid non-linear variance to the judges' decision processes, it should perform worse than the judge: to the extent that it eliminates the random error component in human judgements, it should perform better than the judge.

What were the results of this research? The overwhelming conclusion was that the linear model of the judge's behaviour outperformed the judge. Dawes (1975) noted:

> I know of no studies in which human judges have been able to improve upon optimal statistical prediction . . . A mathematical model by its very nature is an abstraction of the process it models: hence if the decision-maker's behaviour involves following valid principles but following them poorly these valid principles will be abstracted by the model.

Goldberg (1965) reported an intensive study of clinical judgement, pitting experienced and inexperienced clinicians against linear models and a variety of non-linear or configural models in the psychotic/neurotic prediction task. He was led to conclude that Meehl chose the wrong task for testing the clinicians' purported ability to utilize complex configural relationships. The clinicians achieved a 62 per cent hit rate, while the simple linear composite achieved 70 per cent. A 50 per cent hit rate could have been achieved by chance as the criterion base rate was approximately 50 per cent neurotic, 50 per cent psychotic.

Dawes and Corrigan (1974) have called the replacement of the decision-maker by his model *bootstrapping*. Belief in the efficacy of bootstrapping is based on a comparison of the validity of the linear model of the judge with the validity of his or her holistic judgements. However, as Dawes and Corrigan point out, that is only one of two logically possible comparisons. The other is between the validity of the linear model of the judge and the validity of linear models in general. That is, to demonstrate that bootstrapping works because the linear model catches the essence of a judge's expertise and at the same time eliminates unreliability, it is necessary to demonstrate that the weights obtained from an analysis of the judge's behaviour are superior to those that might be obtained in another way – for example, obtained randomly.

Dawes and Corrigan constructed random linear models to predict the criterion. The sign of each predictor variable was determined on an *a priori* basis so that it would have a positive relationship to the criterion.

On average, correlations between the criterion and the output predicted from the random models were *higher* than those obtained from the judge's models. Dawes and Corrigan also investigated equal weighting and discovered that such weighting was even better than the model of the judges or the random linear models. In all cases equal weighting was superior to the models based on judges' behaviour.

Dawes and Corrigan concluded that the human decision-maker need specify with very little precision the weightings to be used in the decision – at least in the context studied; what must be specified is the variables to be used in the linear additive model. It is precisely this knowledge of 'what to look for' in reaching a decision that is the province of the expert clinician. The distinction between knowing what to look for and the ability to integrate information is illustrated in a study by Einhorn (1972). Expert doctors coded biopsies of patients with Hodgkin's disease and then made an overall rating of severity. These overall ratings were very poor predictors of survival time, but the *variables* the doctors coded made excellent predictions when used in a linear additive model.

Factor Analysis

In Chapter 4, we introduced the problem of interpreting large correlation matrices. Take a look at the one set out in Table 11.2. Imagine that the sample size is such that only correlations of ± 0·35 or above are significantly different from zero. What sense can you make of the intercorrelation matrix? Here's a clue: look for *clusters* of highly correlated variables. Can you find any?

Table 11.2. A large correlation matrix

	1	2	3	4	5	6	7	8	9
1 Height	1·00	0·63	0·70	0·01	0·13	0·11	0·15	0·16	0·22
2 Weight		1·00	0·57	0·19	0·09	0·23	0·24	0·19	0·26
3 Shoe size			1·00	0·01	0·17	0·10	0·08	0·01	0·17
4 Talkativeness				1·00	0·61	0·57	0·18	0·30	0·04
5 Number of close friends					1·00	0·47	−0·37	0·10	−0·05
6 Public speaking ability						1·00	0·01	0·21	0·14
7 IQ score							1·00	0·36	0·53
8 Creativity								1·00	0·44
9 Statistical ability									1·00

Let us do the work for you! Look at Table 11.3.

Table 11.3. Correlation matrix with clusters of highly correlated variables indicated

	1	2	3	4	5	6	7	8	9
1 Height	1·00	0·63	0·70	0·01	0·13	0·11	0·15	0·16	0·22
2 Weight		1·00	0·57	0·19	0·09	0·23	0·24	0·19	0·26
3 Shoe size			1·00	0·01	0·17	0·10	0·08	0·01	0·17
4 Talkativeness				1·00	0·61	0·57	0·18	0·30	0·04
5 Number of close friends					1·00	0·47	−0·37	0·10	−0·05
6 Public speaking ability						1·00	0·01	0·21	0·14
7 IQ score							1·00	0·36	0·53
8 Creativity								1·00	0·44
9 Statistical ability									1·00

Height, weight and shoe size seem to form a cluster of intercorrelations. Perhaps they are all aspects of one factor called *bigness*. Similarly, IQ, creativity and statistical ability form a cluster. Perhaps these variables all measure *general intelligence*. Talkativeness, number of close friends and public speaking ability also cluster, suggesting that what we are measuring here are aspects of *extraversion*. The distinctiveness of this cluster of intercorrelations is not as clear-cut as for the other two clusters, however, since IQ score, a variable from the second cluster we identified, also shows a significant negative correlation with number of close friends.

Well, we have just completed an *intuitive factor analysis* of the intercorrelation matrix! Factors are defined as *hypothetical* underlying constructs which we develop to account for obtained intercorrelations between variables. The main objective of factor analysis is to provide a small number of factor constructs that will replace a much larger number of variables. As a researcher you may use factor analysis to test a theory about the number and nature of the factor constructs needed to account for the intercorrelation between the variables that you are studying. For instance the matrix may be the intercorrelation between the, say, thirty items of a questionnaire purportedly measuring 'aggressiveness'. If you obtained two clusters of intercorrelations, the first between items concerned with aggressiveness towards one's own family and friends and the second between items concerned with aggressiveness to people in authority, then you would have strong evidence that aggressiveness is a two-factored disposition. If you are a cross-cultural psychologist you may be interested to discover whether factors related to, say, children's scholastic abilities in mathematics, their native language, foreign language and arts and crafts were stable from culture to culture despite cultural differences in the content of lessons taught. Such questions can be answered by the techniques of factor analysis.

Mathematical factor analysis is preferable to intuitive factor analysis. One major reason for this is that most intercorrelation matrices are not as readily interpretable as the one we have just analysed. Imagine that the positions of variables 2, 5 and 8 in Table 11.3 were interchanged. The resulting matrix would be mathematically similar, but our easily identified clusters would be lost and an intuitive analysis would be difficult. With 100 variables in a matrix, intuitive factor analysis would be near to impossible!

Mathematical factor analysis also allows us to specify *numerically* the contribution each variable in the intercorrelation matrix makes to each factor. The numerical contributions are called *loadings* and they can be thought of as correlations between the variables and the factor.* In Table 11.4 we present a factor analysis of our intercorrelation matrix. Since the loading of height on Factor I is 0·86 and the loading of weight on Factor I is 0·68, it follows that the product of these two loadings is the correlation between weight and height due to Factor I above: $0{\cdot}86 \times 0{\cdot}68 = 0{\cdot}58$.

Now, the correlation between height and weight on our original intercorrelation matrix is 0·63. If we subtract 0·58 from 0·63 this result will be the *residual* correlation between weight and height *after* extraction of Factor I. If all the residuals were as small as this it would be unnecessary to extract any further factors to account for the intercorrelation matrix. Intuitively you should see that we now have the beginnings of one method to help us decide how many factors are necessary to account for an intercorrelation matrix. After each factor is extracted from the matrix a residual correlation matrix is left which contains correlations of diminished absolute size.

* At this point you may have intuitively come to realize that multiple linear regression and factor analysis have similarities. In fact, the basic factor analytic model is also linear and both regression coefficients and factor loadings are weights. In factor analysis, however, there is no criterion variable and analysis is solely concerned with a description of the structure underlying a complex pattern of relationships.

Table 11.4. A factor analysis of the intercorrelation in matrix shown in Table 11.3

	Factors I 'bigness'	II 'extraversion'	III 'general intelligence'
Height	0·86	0·03	0·11
Weight	0·68	0·16	0·24
Shoe size	0·82	0·03	0·00
Talkativeness	−0·01	0·86	0·06
Number of close friends	0·15	0·74	−0·22
Public speaking ability	0·09	0·62	0·16
IQ score	0·09	−0·26	0·79
Creativity	0·04	0·29	0·56
Statistical ability	0·17	0·03	0·65

1 Significance of factor loadings

We have to decide which factor loadings are worth considering when it comes to interpreting factors. As a rule of thumb, loadings to be used in interpretations of the factor matrix should be greater than ± 0·3 and these loadings should be based on an original sample size of *at least* fifty people. Since factor loadings are in effect correlation coefficients it is acceptable to use Appendix Table II in the same way as we tested the significance of Pearson's *r*. This method makes no adjustment for the number of variables involved in a factor matrix, however, and so other significance tests have been proposed. For a more advanced treatment of methods for determining the significance of factor loadings see Child (1977).

2 Factor rotations

Unfortunately, patterns of factor loadings seldom resemble the one we have just described, at least initially! Most methods of factor extraction are designed to extract as much variance as possible from the intercorrelation matrix as each factor is extracted. A much more typical pattern of loadings is shown in Table 11.5. For reasons we will explain later this is called an *unrotated* factor matrix. It is based upon a *different* intercorrelation matrix to the one we have been using so far in this chapter. The underlying intercorrelation matrix contains ten variables.

Table 11.5. An unrotated factor analysis for ten variables (adapted from Comrey, 1973, p. 8)

Variables / Factors	I	II	III	IV	h^2
1	0·48	−0·67	−0·01	−0·05	0·68
2	0·38	−0·63	0·12	−0·08	0·56
3	0·40	−0·65	−0·14	−0·10	0·61
4	0·50	0·27	0·36	−0·17	0·50
5	0·61	0·26	0·37	−0·02	0·57
6	0·46	0·22	0·46	0·09	0·48
7	0·41	0·28	−0·11	−0·40	0·41
8	0·55	0·28	−0·26	−0·25	0·52
9	0·41	0·31	−0·42	0·32	0·54
10	0·47	0·37	−0·38	0·38	0·65
Sum of squares of factor loadings	2·24	1·82	0·91	0·53	5·52

Here, the column h^2 represents what is called the *communality* for each of the variables. It is the sum of squares of the factor loadings for each variable and can take a value as high as 1 or as low as 0. If the communality for a variable were 0 this would indicate that the variable shared nothing in common with the factors, in this case four, that have been extracted. The communality of a variable is the amount of variance in that variable accounted for by the number of factors extracted.

Now if we look at the sum of squares of factor loadings for *each factor* (the last row) we can see how much of total variance, in this case ten because there are ten variables, each successive factor accounts for. Factor I is the largest factor in that it accounts for 22·4 per cent of the variance. Altogether the four factors account for 22·4 + 18·2 + 9·1 + 5·3 = 55 per cent of the variance in the intercorrelation matrix.

However, factor analysis does not stop with the extraction of an *unrotated* factor matrix where the first factor extracted accounts for the major share of variance in the intercorrelation matrix. This is because unrotated matrices are very difficult to interpret since they often involve complex overlapping factors. For this reason factor matrices are often *rotated* to enable the researcher to interpret the factors. Alternative rotations are mathematically equivalent.

What is an unrotated factor matrix and what are rotations? Perhaps the best way to explain is to present a graphical representation of the mathematical processes involved. We will use some new data.

Consider Figure 11.1. It is a visual representation of the loadings on Factors I and II of four variables. All four variables have moderate loadings on Factor I. Variables 1 and 2 have moderate negative loadings on Factor II whilst variables 3 and 4 have moderate positive loadings on this factor.

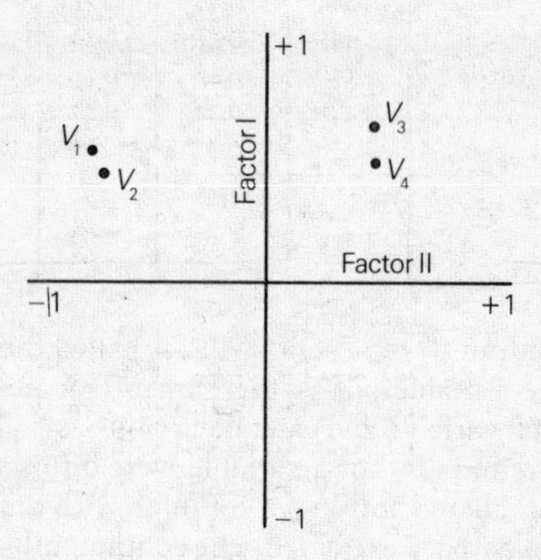

Figure 11.1 A visual representation of an unrotated factor analysis

The factors can be represented by orthogonal lines since the factors themselves are orthogonal or uncorrelated. Now, it is mathematically valid to *rotate* these axes around the origin whilst still maintaining the right angle between the axes representing the two distinct factors. This means we can rotate the axes to the position shown in Figure 11.2.

This rotation results in variables 1 and 2 having moderate loadings on Factor I with no appreciable loadings on Factor II. Similarly

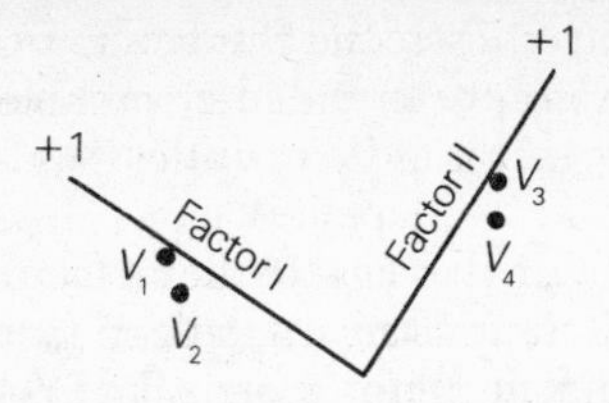

Figure 11.2 A rotated version of the factor analysis shown in Figure 11.1

variables 3 and 4 now have moderate loadings on Factor II with no appreciable loading on Factor I.

Now consider Table 11.6, which is a *varimax* rotated factor analysis of the unrotated analysis we presented in Table 11.5. Varimax rotation is a rotation which attempts to *maximize and minimize* loadings on each factor so that loadings approach 1 and 0.

Table 11.6. A varimax factor analysis corresponding to the unrotated analysis shown in Table 11.5 (adapted from Comrey, 1973, p. 10)

Variables	Factors I	II	III	IV	h^2
1	0·82	0·04	0·08	0·01	0·68
2	0·73	0·10	−0·10	−0·02	0·56
3	0·76	−0·10	0·14	0·03	0·61
4	0·03	0·64	0·28	−0·05	0·50
5	0·09	0·72	0·20	0·08	0·57
6	0·05	0·69	0·00	0·02	0·48
7	−0·02	0·22	0·60	0·06	0·41
8	0·04	0·24	0·60	0·31	0·52
9	−0·06	0·12	0·16	0·70	0·54
10	−0·08	0·21	0·14	0·76	0·65
Sum of squares of factor loadings	1·81	1·59	0·92	1·18	5·52

Notice that the communalities are the same after the varimax rotation, but the amount of variance extracted from the intercorrelation matrix *by each factor* has changed. However, the four factors in total still account for 55 per cent of the total variance (18·1 + 15·9 +

9·2 + 11·8 = 55). You will see that the factor structure is now much clearer. Factor I is mainly determined by variables 1 to 3, Factor II by variables 4, 5 and 6, Factor III by variables 7 and 8 and Factor IV by variables 9 and 10.

There are many different ways to extract factors from the correlation matrix. Indeed, there has been little in factor analytic methodology to say which solution is more valid. You will come across methods of factoring like principal factoring, canonical factoring, alpha factor and the like. Perhaps one of the most promising is *confirmatory maximum likelihood factor analysis*. This method, for the first time, makes it possible to cross-validate factors directly rather than by judgement. The factor structure identified in one sample can be tested on a second sample. This is achieved by restricting, on the basis of the first exploratory analysis, the values that certain factor loadings can take. In the second sample, changes in the unrestricted loadings can then be monitored. The rotations available too are varied. Some allow factors to be themselves correlated. For instance, it would be possible to represent the factors on p. 232 by non-orthogonal axes. Particular methods of factor extraction combined with particular rotations and particular rules for deciding when to stop extracting further factors produce different results *with the same data*. The factor analysis presented in Table 11.5 would seem to suggest a general factor, to which all variables contribute, together with three more specific factors. The factor analysis shown in Table 11.6, on the other hand, suggests four specific factors.

For these reasons we would argue that those theories of intelligence which have been based on factor analytic methodology may have been heavily influenced by the particular methodology that was widely available and used at the time. Some theories emphasize a single general factor (e.g. Spearman), other theorists emphasize a hierarchy of factors (e.g. Burt), some theorists identify several primary factors (e.g. Thurstone), whilst other theorists specify many specific factors (e.g. Guildford).

It follows that factor analysis has been called an art rather than a science. It is impossible to specify the best factor analysis of a correlation matrix, *but* it is possible to reject potential factor structures that do not fit the observed data. A final choice between factoring methods and analyses that are statistically valid must be made on psychological grounds.

Multidimensional Scaling

A problem faced by many researchers is how to measure and understand the way people view relationships between objects. Multidimensional scaling analysis allows us to represent visually the psychological similarities between objects or experiences as points on a scattergram. Decreasing physical separation between objects represents increasing psychological similarity. Objects judged to be psychologically dissimilar will be represented as being far apart. For example, consider Figure 11.3!

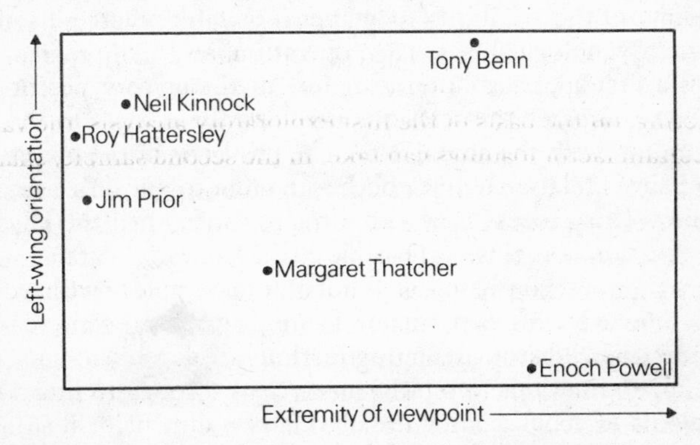

Figure 11.3 A 'possible' multidimensional scaling representation

By asking people to rate the similarity of objects and subsequently representing the similarity as physical distance it is possible to make *interpretations* of the *underlying dimensions* on which the objects have been judged relative to one another – in this case, 'left-wing orientation' and 'extremity of viewpoint'. Multidimensional scaling (MDS) procedures do not require any *a priori* knowledge of these dimensions and, since only similarity judgements are required, the experimenter does not 'put words into the subject's mouth'. Experimenter effects (see Chapter 2) can, for this reason, be expected to be negligible – the role of language is kept to a minimum.

There is a fundamental distinction between factor analysis and multidimensional scaling. Whilst the former is principally concerned with data reduction, in order to make a large data set more manageable, MDS is an attempt to measure and understand the relations between objects *within an assumed spatial model of psychological similarity*. MDS is simply a mathematical tool for representing the adjudged similarities between objects as a spatial map.

In multidimensional scaling studies, each possible pairing of objects from an object set is presented to the respondent who then rates the similarity of the pair on a seven-point scale ranging from, say, no similarity at all to completely similar. In most situations these ratings will be *ordinal-level*. The intention of *non-metric* multidimensional scaling is to represent, as closely as possible, the rank order of the similarity judgements as rank-ordered distances in some psychological 'space'. In operation, an MDS program calculates a first approximation to represent the objects' positioning and then compares the relations between the objects in this approximation with the actual data. The positions of the objects are then changed relative to one another in order to minimize any discrepancy. This process is repeated many times until the physical representation is as close as possible to the similarity data.

Fitting the best configuration is not quite as simple as we have just stated, however, MDS programs cannot evaluate *every* possible configuration. Subsequent approximations are dependent upon previous approximations as a program iterates through to find a best possible fit by repositioning the objects in *n*-dimensional space to locate the next best fit, the so-called *method of steepest descent*. For this reason, a program's 'best-fitting' solution may be a 'local minima' rather than the *best possible fit*. This problem is often solved by starting the program on several different first approximations and so obtaining several different solutions. If the solutions converge it is very likely that an obtained 'best fit' is in reality *the* best possible fit to the data. Often the best representation is not two-dimensional, as in a scattergram, but may contain several dimensions. Two, three or four dimensions may be needed to represent best the structure of the similarity data. While it is impossible to visualize more than three dimensions, there is no difficulty in representing extra dimensions mathematically and the 'higher' dimensions are conventionally displayed by plotting them with respect to the first or second dimension.

1 How many dimensions?

One guide to deciding how many dimensions are necessary to account for a particular set of similarity data is to compute the correlation between similarity rankings and the distance rankings produced by a particular *n*-dimensional representation. In general, 'goodness of fit' will increase as the number of dimensions increases. For this reason methods have been developed which monitor increments in goodness of fit as dimensionality increases. If the increment in goodness of fit achieved by adding a further dimension is relatively small then the extra dimension is generally not considered.

The final decision on number of dimensions should be *interpretability*. Computer programs which perform multidimensional scaling analysis plot two-dimensional scattergrams of each pairing of dimensions in the *n*-dimensional solution. It is then up to you to interpret the scattergrams. As in the case of factors, the interpretation and naming of dimensions has been described as an art rather than a science. Most researchers adopt the rule of thumb that if a dimension does not lend itself to an easy interpretation, then it probably does not exist. Several strategies can help you interpret dimensions. The first is to look at the objects represented at the ends of the dimension and see if there is some attribute that changes in an obvious way. The second is to look for 'clusters' of objects, as clustering implies some shared similarity.

One major advantage of multidimensional scaling over other multivariate techniques is that it can be used with data obtained from only a few subjects. Indeed, stable representations can be obtained concerning the perceptions of a single subject. Since the process of eliciting paired comparisons data can be very time-consuming for the investigator and boring for the investigator's subject, many multidimensional scaling programs now allow you to input an incomplete set of paired comparisons. The program then estimates the missing data. Also it is possible to analyse data from other ranking and grouping tasks which can be made more interesting to the subject.

2 Uses of multidimensional scaling: an illustrative example

In some cases, it can be shown that MDS techniques can be used to elicit subjective perceptions that are *not accessible* with the questionnaire methodologies. Questionnaires are dependent upon the mediating role of language and may therefore be inappropriate for some situations. Whalley (1984) details just such a case, involving the perception of pain.

The dominant methodology at present used to explore pain perception involves checking descriptive words on a questionnaire. From the words an individual has checked to describe his pain it is possible to make a *preliminary* diagnosis. This diagnosis is based on previously observed relations between words checked and *subsequent* final diagnoses, based in some cases on surgical investigation. Words included in the questionnaire include 'pulsing', 'drilling', 'cutting', 'tender', 'dull', 'tight' etc. Whalley showed that these linguistic descriptions have different meanings for different people and advocates a multidimensional scaling approach to pain. In his study, patients were required to make comparative similarity judgements between the pain that they were experiencing and a set of

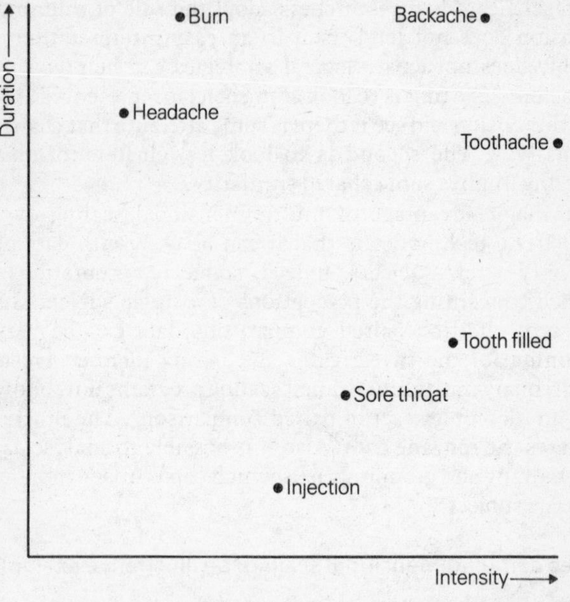

Figure 11.4 A multidimensional scaling approach to pain perception (after Whalley, 1984)

commonly occurring painful events. The relational information shown between the new pain and the set of common pains can then be used as a method for diagnosis without requiring patients to use imprecise ambiguous adjectives. One of Whalley's two-dimensional reference plots is given in Figure 11.4. He has labelled the two pain dimensions 'intensity' and 'duration'.

Summary

This chapter detailed the rationale and use of several multivariate techniques. Partial correlation measures the relationship between two variables whilst removing the effect of a third variable. Multiple linear regression evaluates the relationship between multiple predictor variables and a single criterion. Factor analysis simplifies and aids interpretation of the information contained in large intercorrelation matrices. Finally, multidimensional scaling represents visually the psychological similarity of objects or experiences.

12 Investigative Design and Statistics: a Reconciliation and Conclusion

In this chapter we attempt to reconcile experimental and correlational designs by adopting two approaches: first, a statistical approach illustrating the essential similarity of the results of a *t*-test and a correlational analysis of the same data set and, second, an empirical approach outlining the usefulness of analysis of variance methodology in the domain of personality psychology. We end with an overview of investigative design and statistics in psychology.

Experimental and Correlational Designs: a Statistical Reconciliation

Experiments are concerned with locating the cause of *differences* between two or more conditions. The *independent* variable is manipulated by the experimenter and has at least two categories – the presence and the absence of a treatment. The *dependent* variable is a measure of subjects' performance under each of the two conditions. The *t*-test is often the appropriate method of analysis. If the independent variable has more than two categories, one-way analysis of variance is often used. If several independent variables are involved more elaborate analyses of variance techniques are appropriate.

However, the results of such experiments can be analysed using correlational techniques. Imagine an experiment to investigate the usefulness of training in mnemonic techniques to aid recall of a list of sixty 'unrelated' words. Ten subjects attempted to recall the whole list without any prior training and the number of words correctly recalled is noted. Ten other subjects were given a half-hour's training in the use of mnemonic techniques before being shown the list of words and then performing the recall task. The resulting data are set out below:

Recall without training in mnemonic techniques	Recall with training in mnemonic techniques
40	50
15	44
17	65
30	45
25	51
28	70
43	60
27	48
33	10
31	58

With only two conditions the significance of the difference between the means may be tested with a *t*-test or a one-way analysis of variance. The two procedures lead to the same result when the obtained *t*- and *F*-values are referred to the significance tables.

In fact, $\sqrt{F} = t$. Since the mathematical proof is long-winded we will not give it here.

For our particular imaginary experiment:

$F = 11{\cdot}90$ (following the step-by-step procedure from Chapter 9) and $t = -3{\cdot}45$.

Both the obtained *F*- and *t*-values are significant at the 0·01 level for a one-tailed test.

Now, instead of performing an analysis of variance or a *t*-test on the data we could calculate the *point-biserial correlation coefficient*. This correlation coefficient is, like Spearman's rho, a special case of the product-moment correlation coefficient. It is a coefficient to measure the correlation between a dichotomous *nominal-level variable* and an interval-level variable. In our case, the nominal-level variable is the presence or absence of mnemonic training. We will assign the absence of mnemonic training a numerical label of 1 and we will give a numerical label of 2 to the presence of mnemonic training. These two labels are arbitrary. We could have used 10 and 15 – it would make no difference to our calculations.

Let us lay out this transformation of our experimental data:

1	1	1	1	1	1	1	1	1	1	2	2	2	2	2	2	2	2	2	2
40	15	17	30	25	28	43	27	33	31	50	44	65	45	51	70	60	48	10	58

If we follow the step-by-step procedure for the product-moment correlation coefficient from Chapter 6, the point-biserial correlation works out to be 0·64. This point-biserial correlation is significant at the 0·01 level for a one-tailed test. Notice that this result is at the same level of significance as the *F*-ratio and the *t*-value we obtained from the untransformed data. In this situation, testing the null hypothesis that $\mu_1 = \mu_2$ is identical to testing the null hypothesis that $\varrho = 0$.

So, a problem involving the comparison of two mean scores can be converted to a problem of estimating the strength of association between two variables. Methods for converting comparisons of more than two means to a correlational analysis are more complex but theoretically similar. These methods use multiple correlation and multiple regression. Indeed it can be shown that multiple regression is a general method that subsumes analysis of variance!

Should multiple regression analysis replace analysis of variance? Kerlinger (1973) argues that it should not, at least for teaching purposes. The structuring and partitioning of data, a necessary prerequisite before the computation of ANOVA, serve a useful pedagogical purpose. Also, multiple regression involves complex mathematics – beyond the scope of a step-by-step cookbook!

Perhaps one major reason why the relationship between experimental and correlational statistical techniques has received little attention in books on statistics for social scientists is linked to Lee Cronbach's analysis of the divide between psychologists working within the experimental research format and psychologists working within the correlational research format, as we saw in Chapter 4. Concern with measures of association and correlation, regression, factor analysis and multidimensional scaling are viewed as the domain of the correlational psychologist, while *t*-tests and analysis of variance are seen as typifying the experimental psychologist.

Recently, however, the methods of analysis of variance have been used with some success in personality psychology. In the next section of this chapter we present an overview of the background to this meeting of the 'two disciplines of scientific psychology'.

Personality Psychology and Analysis of Variance

A major topic in the study of personality is a concern with identifying the determinants of behaviour in situations. Early personality research was dominated by trait and psycho-dynamic theories which *assumed* the existence of trans-situational consistency (e.g. Cattell, 1946; Guildford, 1959; McClelland, 1951). In other words, stable characteristics of an individual were seen to determine the way that that individual behaves in any, or most, situations. This conceptualization of the individual and his interaction with the environment has been labelled *personologism* (e.g. Adinolfi, 1971). Recent cognitive style research can also be labelled in this way (e.g. Witkin, 1962).

Later, social learning theorists emphasized the importance of the situation as a determinant of the way an individual behaves. In the domain of decision-making, Payne's (1982) review has emphasized that decision behaviour is highly contingent on task characteristics. This focus on the situation as the main source of behavioural variation has been labelled *situationism* (e.g. Mischel, 1968).

However, as Endler (1975) pointed out, 'situations do not exist in a vacuum but have psychological meaning and significance for people'. People select, create and construct their own psychological environments. These arguments have led to a study of the *interaction* between persons and situations in an attempt to identify the locus of behavioural variation. Bowers (1973) has noted that almost all recent studies investigating the source of behavioural variation have concluded that interactionism is more important than either personologism or situationism. In short, interactionism would seem to be the major contemporary conceptualization of personality.

Two major methodologies have been used for investigation of the relative contribution of the person, the situation or an interaction of the two. The first strategy has been simply to correlate measures of a personality trait or behaviour. As Endler (1975) pointed out, this strategy has usually yielded correlations of 0·30 and such results have usually been taken to support the situationist position. However, it must be noted that whilst such low correlations do not support a personologist or trait position, they do not differentiate between situationism and interactionism. This is because correlations may be attenuated, or lowered, by interactions, in addition to situational specificity.

The second, more recent, research strategy has been to use an analysis of variance approach. This allows the relative variances contributed by situation, persons and an interaction of these to be evaluated. Endler and Hunt (1968) provided one of the first demonstrations of this technique in an investigation of the person–situation issue. In essence, the development of the ANOVA approach has made possible comparisons between the personologist, situationist and interactionist positions. Endler (1966) presents an account of a variance components technique that surmounts the methodological problem of directly comparing mean squares from different sources of variance where the mean squares are not independent of one another.

However, there are some difficulties with the ANOVA approach. Recently some personality psychologists have pointed to some potential problems with what would appear, at first sight, to be an ideal methodology for disentangling situationist, interactionist and personologist accounts of behavioural variance. Olweus (1977) has noted that even if a large situation-by-person interaction variance is found it may have arisen in many different ways. He argues that it is impossible to make a clear empirical test of the interactionist position. Nisbett (1977) has also argued that the major disadvantage of interaction hypotheses is that they are much more difficult to refute than main-effect hypotheses. Also, interactions involving several levels on each of the independent variables can be extremely difficult to describe and comprehend:

> It is far from uncommon to discover that the results of a . . . design are virtually uninterpretable, because statements about main effects or interactions must be hedged around and constantly modified, in thought and communication, by qualifications necessitated by the presence of higher order interactions. (Nisbett, 1977, p. 241)

In other words, Nisbett is arguing that most reports of interactions are often post-hoc.

Another limitation with the analysis of variance approach would seem to be that the selective sampling of situations and persons can, perhaps unintentionally, alter the relative magnitudes of the variance components.

Golding (1975) has been critical of the use of the ANOVA methodology. He argues that generalizability theory (Cronbach et al., 1972) provides more appropriate statistics. The relative sizes of interaction terms must be validated by showing that they are non-artificial, replicable and meaningfully patterned. Perhaps a more acute problem is that ANOVA gives only an indication of *how much* the variance components contribute to behavioural variation. Deeper theoretical questions about *how* situations and individuals interact to produce observed behaviour are not answered directly by the statistics of ANOVA. Ayton and Wright (in press) discuss these issues and detail an improved methodology with which to approach these problems.

Currently, taxonomies of situations, necessary for the best application of ANOVA, are little developed compared to the multiplicity of personality types that have been described. However, pairwise similarity scaling of situations, input into a multidimensional scaling analysis, would seem to offer promise in establishing a taxonomy of situations based on individuals' perceptions. Other multivariate techniques, such as factor analysis, could also be used much more extensively to aid the systematic development of taxonomies of traits (or styles) and situations.

Conclusion

Having read this book you should now have an understanding of the research process as it takes place within psychology. You should appreciate the interdependence between all stages of the research process and realize that decisions taken at any stage place restrictions on subsequent decision-making. Your own theoretical perspective can be seen as a major force underlying the way you carry out your own research. Nevertheless, political and social influences can be shown to have a considerable impact on research design and procedure and also on data analysis and interpretation. Your awareness of these influences should enable you to take a broader perspective when analysing and evaluating the types of psychological research that have been undertaken.

You should also have an appreciation of how levels of measurement and measurement strategy influence the appropriateness of choice between statistical tests. Your choice of research format, data-collection technique and measurement scale will all limit the range of statistical tests available to you to analyse the data that you have collected. Your choice of research format, correlational or experimental, will also constrain your ability to draw causal inferences. Although a well-controlled experiment provides the best conditions for inferring causation its result may not generalize to situations and people outside the psychological laboratory.*

We also discussed the concepts and assumptions underlying statistical testing and saw that parametric tests have certain advantages over their non-parametric equivalents.

In terms of the statistical techniques underlying the correlational research format, you should appreciate the advantages of measuring strength of association rather than independence between two nominal-level variables. You should also understand the rationale and calculations of Goodman–Kruskal's lambda, Pearson's *r* and Spearman's rho. You should be able to evaluate the significance of an obtained correlation coefficient, compute confidence intervals and also compare the difference between two correlation coefficients for independent samples. You should know how to fit a least-squares regression line and understand the close relationship between prediction and correlation. You should see how correlation and regression form a background for questionnaire design and analysis, and appreciate the crucial role of reliability and validity evaluation in questionnaire design.

As we saw, analysis of variance plays a major role in the experimental research format, making it possible to isolate the contributions of several independent variables to changes in a single dependent variable. Parametric analysis of variance also allows the possibility of interactive effects, produced by two or more independent variables, to be investigated. This possibility exists in a simple form only for non-parametric data, illustrating the important constraint of level of measurement on the possibilities of data analysis.

You should be aware of the potential usefulness of multivariate techniques in data analysis. Although the computations underlying these techniques are complex and far beyond the scope of step-by-step formulae, multivariate analysis is available on main-frame computers in the form of statistical packages such as SPSS and BMDP.

* We recommend two statistical packages to aid data analysis. Both are available for Apple II and BBC microcomputers. The first, SUPASTAT, covers most of the tests detailed in Chapters 6, 7, 8 and 9 of this book. The second, OMNIBUS, deals with non-parametric techniques. For details of SUPASTAT (version 2.2P) contact R. Eglen, 17 Tanhouse Park, Hippoholme, Halifax. Details of OMNIBUS can be found in R. Meddis's book *Statistics using ranks*, published by Blackwell in 1984. Alternatively, contact R. Meddis at the Department of Human Sciences, University of Technology, Loughborough, Leicestershire.

These two programs are commonly available in most university and polytechnic psychology departments. However, each requires some fairly extensive initial training in its use. In the near future, multivariate analysis will become readily available on microcomputers in a form such that the user with no previous experience of the program will be able to perform the types of analysis sketched out here.

Finally, you should now be aware that experimental and correlational statistical techniques lead to essentially the same results when used to analyse the same set of data. Problems involving the comparison of means can be converted to problems of estimating the strength of relationships between variables. Use of analysis of variance methodology in the traditional correlational domain of personality psychology has aided the development of psychological theory. For these reasons, our initial strong distinction between the experimental and correlational research formats has become blurred. The statistics associated with one research format *can* be used in the other. As we saw in Chapter 4, the major distinction between the experimental and correlational psychologist is in the use of active versus passive observation.

Appendix

List of Statistical Tables

Table I Proportions of area under the normal distribution curve

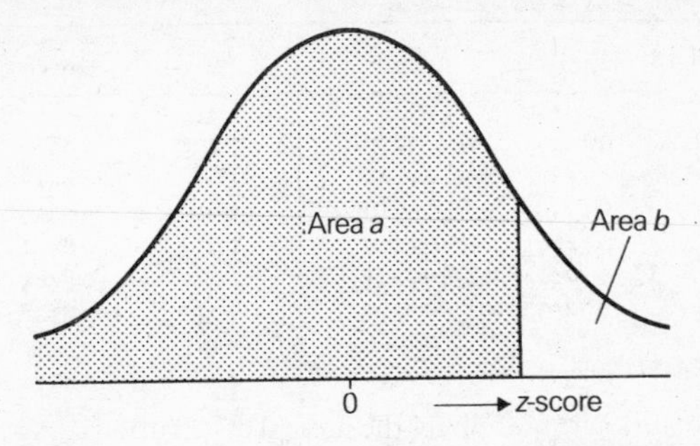

Note that the area under the curve can be expressed as a probability – so the critical Z for $P < 0{\cdot}05$ for the two areas is underlined.

z	area *a*	area *b*	*z*	area *a*	area *b*
0·00	0·5000	0·5000	0·28	0·6103	0·3897
0·01	0·5040	0·4960	0·29	0·6141	0·3859
0·02	0·5080	0·4920	0·30	0·6179	0·3821
0·03	0·5120	0·4880	0·31	0·6217	0·3783
0·04	0·5160	0·4840	0·32	0·6255	0·3745
0·05	0·5199	0·4801	0·33	0·6293	0·3707
0·06	0·5239	0·4761	0·34	0·6331	0·3669
0·07	0·5279	0·4721	0·35	0·6368	0·3632
0·08	0·5319	0·4681	0·36	0·6406	0·3594
0·09	0·5359	0·4641	0·37	0·6443	0·3557
0·10	0·5398	0·4602	0·38	0·6480	0·3520
0·11	0·5438	0·4562	0·39	0·6517	0·3483
0·12	0·5478	0·4522	0·40	0·6554	0·3446
0·13	0·5517	0·4483	0·41	0·6591	0·3409
0·14	0·5557	0·4443	0·42	0·6628	0·3372
0·15	0·5596	0·4404	0·43	0·6664	0·3336
0·16	0·5636	0·4364	0·44	0·6700	0·3300
0·17	0·5675	0·4325	0·45	0·6736	0·3264
0·18	0·5714	0·4286	0·46	0·6772	0·3228
0·19	0·5753	0·4247	0·47	0·6808	0·3192
0·20	0·5793	0·4207	0·48	0·6844	0·3156
0·21	0·5832	0·4168	0·49	0·6879	0·3121
0·22	0·5871	0·4129	0·50	0·6915	0·3085
0·23	0·5910	0·4090	0·51	0·6950	0·3050
0·24	0·5948	0·4052	0·52	0·6985	0·3015
0·25	0·5987	0·4013	0·53	0·7019	0·2981
0·26	0·6026	0·3974	0·54	0·7054	0·2946
0·27	0·6064	0·3936	0·55	0·7088	0·2912

Table I Proportions of area under the normal distribution curve (continued)

z	area a	area b	z	area a	area b
0·56	0·7123	0·2877	0·99	0·8389	0·1611
0·57	0·7157	0·2843	1·00	0·8413	0·1587
0·58	0·7190	0·2810	1·01	0·8438	0·1562
0·59	0·7224	0·2776	1·02	0·8461	0·1539
0·60	0·7257	0·2743	1·03	0·8485	0·1515
0·61	0·7291	0·2709	1·04	0·8508	0·1492
0·62	0·7324	0·2676	1·05	0·8531	0·1469
0·63	0·7357	0·2643	1·06	0·8554	0·1446
0·64	0·7389	0·2611	1·07	0·8577	0·1423
0·65	0·7422	0·2578	1·08	0·8599	0·1401
0·66	0·7454	0·2546	1·09	0·8621	0·1379
0·67	0·7486	0·2514	1·10	0·8643	0·1357
0·68	0·7517	0·2483	1·11	0·8665	0·1335
0·69	0·7549	0·2451	1·12	0·8686	0·1314
0·70	0·7580	0·2420	1·13	0·8708	0·1292
0·71	0·7611	0·2389	1·14	0·8729	0·1271
0·72	0·7642	0·2358	1·15	0·8749	0·1251
0·73	0·7673	0·2327	1·16	0·8770	0·1230
0·74	0·7704	0·2296	1·17	0·8790	0·1210
0·75	0·7734	0·2266	1·18	0·8810	0·1190
0·76	0·7764	0·2236	1·19	0·8830	0·1170
0·77	0·7794	0·2206	1·20	0·8849	0·1151
0·78	0·7823	0·2177	1·21	0·8869	0·1131
0·79	0·7852	0·2148	1·22	0·8888	0·1112
0·80	0·7881	0·2119	1·23	0·8907	0·1093
0·81	0·7910	0·2090	1·24	0·8925	0·1075
0·82	0·7939	0·2061	1·25	0·8944	0·1056
0·83	0·7967	0·2033	1·26	0·8962	0·1038
0·84	0·7995	0·2005	1·27	0·8980	0·1020
0·85	0·8023	0·1977	1·28	0·8997	0·1003
0·86	0·8051	0·1949	1·29	0·9015	0·0985
0·87	0·8078	0·1922	1·30	0·9032	0·0968
0·88	0·8106	0·1894	1·31	0·9049	0·0951
0·89	0·8133	0·1867	1·32	0·9066	0·0934
0·90	0·8159	0·1841	1·33	0·9082	0·0918
0·91	0·8186	0·1814	1·34	0·9099	0·0901
0·92	0·8212	0·1788	1·35	0·9115	0·0885
0·93	0·8238	0·1762	1·36	0·9131	0·0869
0·94	0·8264	0·1736	1·37	0·9147	0·0853
0·95	0·8289	0·1711	1·38	0·9162	0·0838
0·96	0·8315	0·1685	1·39	0·9177	0·0823
0·97	0·8340	0·1660	1·40	0·9192	0·0808
0·98	0·8365	0·1635	1·41	0·9207	0·0793

Table I Proportions of area under the normal distribution curve (continued)

z	area a	area b	z	area a	area b
1·42	0·9222	0·0778	1·85	0·9678	0·0322
1·43	0·9236	0·0764	1·86	0·9686	0·0314
1·44	0·9251	0·0749	1·87	0·9693	0·0307
1·45	0·9265	0·0735	1·88	0·9699	0·0301
1·46	0·9279	0·0721	1·89	0·9706	0·0294
1·47	0·9292	0·0708	1·90	0·9713	0·0287
1·48	0·9306	0·0694	1·91	0·9719	0·0281
1·49	0·9319	0·0681	1·92	0·9726	0·0274
1·50	0·9332	0·0668	1·93	0·9732	0·0268
1·51	0·9345	0·0655	1·94	0·9738	0·0262
1·52	0·9357	0·0643	1·95	0·9744	0·0256
1·53	0·9370	0·0630	1·96	0·9750	0·0250
1·54	0·9382	0·0618	1·97	0·9756	0·0244
1·55	0·9394	0·0606	1·98	0·9761	0·0239
1·56	0·9406	0·0594	1·99	0·9767	0·0233
1·57	0·9418	0·0582	2·00	0·9772	0·0228
1·58	0·9429	0·0571	2·01	0·9778	0·0222
1·59	0·9441	0·0559	2·02	0·9783	0·0217
1·60	0·9452	0·0548	2·03	0·9788	0·0212
1·61	0·9463	0·0537	2·04	0·9793	0·0207
1·62	0·9474	0·0526	2·05	0·9798	0·0202
1·63	0·9484	0·0516	2·06	0·9803	0·0197
1·64	0·9495	0·0505	2·07	0·9808	0·0192
1·65	0·9505	0·0495	2·08	0·9812	0·0188
1·66	0·9515	0·0485	2·09	0·9817	0·0183
1·67	0·9525	0·0475	2·10	0·9821	0·0179
1·68	0·9535	0·0465	2·11	0·9826	0·0174
1·69	0·9545	0·0455	2·12	0·9830	0·0170
1·70	0·9554	0·0446	2·13	0·9834	0·0166
1·71	0·9564	0·0436	2·14	0·9838	0·0162
1·72	0·9573	0·0427	2·15	0·9842	0·0158
1·73	0·9582	0·0418	2·16	0·9846	0·0154
1·74	0·9591	0·0409	2·17	0·9850	0·0150
1·75	0·9599	0·0401	2·18	0·9854	0·0146
1·76	0·9608	0·0392	2·19	0·9857	0·0143
1·77	0·9616	0·0384	2·20	0·9861	0·0139
1·78	0·9625	0·0375	2·21	0·9864	0·0136
1·79	0·9633	0·0367	2·22	0·9868	0·0132
1·80	0·9641	0·0359	2·23	0·9871	0·0129
1·81	0·9649	0·0351	2·24	0·9875	0·0125
1·82	0·9656	0·0344	2·25	0·9878	0·0122
1·83	0·9664	0·0336	2·26	0·9881	0·0119
1·84	0·9671	0·0329	2·27	0·9884	0·0116

Table I Proportions of area under the normal distribution curve (continued)

z	area a	area b	z	area a	area b
2·28	0·9887	0·0113	2·71	0·9966	0·0034
2·29	0·9890	0·0110	2·72	0·9967	0·0033
2·30	0·9893	0·0107	2·73	0·9968	0·0032
2·31	0·9896	0·0104	2·74	0·9969	0·0031
2·32	0·9898	0·0102	2·75	0·9970	0·0030
2·33	0·9901	0·0099	2·76	0·9971	0·0029
2·34	0·9904	0·0096	2·77	0·9972	0·0028
2·35	0·9906	0·0094	2·78	0·9973	0·0027
2·36	0·9909	0·0091	2·79	0·9974	0·0026
2·37	0·9911	0·0089	2·80	0·9974	0·0026
2·38	0·9913	0·0087	2·81	0·9975	0·0025
2·39	0·9916	0·0084	2·82	0·9976	0·0024
2·40	0·9918	0·0082	2·83	0·9977	0·0023
2·41	0·9920	0·0080	2·84	0·9977	0·0023
2·42	0·9922	0·0078	2·85	0·9978	0·0022
2·43	0·9925	0·0075	2·86	0·9979	0·0021
2·44	0·9927	0·0073	2·87	0·9979	0·0021
2·45	0·9929	0·0071	2·88	0·9980	0·0020
2·46	0·9931	0·0069	2·89	0·9981	0·0019
2·47	0·9932	0·0068	2·90	0·9981	0·0019
2·48	0·9934	0·0066	2·91	0·9982	0·0018
2·49	0·9936	0·0064	2·92	0·9982	0·0018
2·50	0·9938	0·0062	2·93	0·9983	0·0017
2·51	0·9940	0·0060	2·94	0·9984	0·0016
2·52	0·9941	0·0059	2·95	0·9984	0·0016
2·53	0·9943	0·0057	2·96	0·9985	0·0015
2·54	0·9945	0·0055	2·97	0·9985	0·0015
2·55	0·9946	0·0054	2·98	0·9986	0·0014
2·56	0·9948	0·0052	2·99	0·9986	0·0014
2·57	0·9949	0·0051	3·00	0·9987	0·0013
2·58	0·9951	0·0049	3·01	0·9987	0·0013
2·59	0·9952	0·0048	3·02	0·9987	0·0013
2·60	0·9953	0·0047	3·03	0·9988	0·0012
2·61	0·9955	0·0045	3·04	0·9988	0·0012
2·62	0·9956	0·0044	3·05	0·9989	0·0011
2·63	0·9957	0·0043	3·06	0·9989	0·0011
2·64	0·9959	0·0041	3·07	0·9989	0·0011
2·65	0·9960	0·0040	3·08	0·9990	0·0010
2·66	0·9961	0·0039	3·09	0·9990	0·0010
2·67	0·9962	0·0038	3·10	0·9990	0·0010
2·68	0·9963	0·0037	3·11	0·9991	0·0009
2·69	0·9964	0·0036	3·12	0·9991	0·0009
2·70	0·9965	0·0035	3·13	0·9991	0·0009

Table I Proportions of area under the normal distribution curve (continued)

z	area a	area b	z	area a	area b
3·14	0·9992	0·0008	3·22	0·9994	0·0006
3·15	0·9992	0·0008	3·23	0·9994	0·0006
3·16	0·9992	0·0008	3·24	0·9994	0·0006
3·17	0·9992	0·0008	3·30	0·9995	0·0005
3·18	0·9993	0·0007	3·40	0·9997	0·0003
3·19	0·9993	0·0007	3·50	0·9998	0·0002
3·20	0·9993	0·0007	3·60	0·9998	0·0002
3·21	0·9993	0·0007	3·70	0·9999	0·0001

Reprinted by permission from Table A in *Fundamentals of Behavioral Sciences* by Runyon and Haber, 3rd edn 1976, copyright © Addison-Wesley, Reading, Mass., USA.

Table II Critical values of *r* (Pearson's correlation coefficient)*

	2-tail		10%	5%	2%	0·2%
	1-tail		5%	2·5%	1%	0·1%
N =	4	r ⩾	0·90	0·95	0·98	0·995
	5		0·81	0·88	0·93	0·97
	6		0·73	0·81	0·88	0·94
	7		0·67	0·75	0·83	0·90
	8		0·62	0·71	0·79	0·87
	9		0·58	0·67	0·75	0·84
	10		0·55	0·63	0·72	0·81
	11		0·52	0·60	0·69	0·78
	12		0·50	0·58	0·66	0·76
	13		0·48	0·55	0·63	0·74
	14		0·46	0·53	0·61	0·72
	15		0·44	0·51	0·59	0·70
	16		0·43	0·50	0·57	0·68
	17		0·41	0·48	0·56	0·66
	18		0·40	0·47	0·54	0·65
	19		0·39	0·46	0·53	0·64
	20		0·38	0·44	0·52	0·62
	21		0·37	0·43	0·50	0·61
	22		0·36	0·42	0·49	0·60
	23		0·35	0·41	0·48	0·59

* Strictly speaking, Spearman's rho coefficient should be referred to a separate table. But the two tables are similar when the correlations are given to two decimal places and when *N* is greater than 10.

Table II Critical values of *r* (Pearson's correlation coefficient) (continued)

2-tail	10%	5%	2%	0·2%
1-tail	5%	2·5%	1%	0·1%
24	0·34	0·40	0·47	0·57
25	0·34	0·40	0·46	0·56
30	0·31	0·36	0·42	0·52
40	0·26	0·30	0·36	0·45
50	0·24	0·28	0·34	0·42
60	0·22	0·26	0·31	0·38
70	0·20	0·24	0·28	0·35
80	0·18	0·22	0·26	0·33
90	0·17	0·21	0·24	0·32
100	0·16	0·20	0·23	0·30

Table III Fisher's transform of Pearson's correlation coefficient

r	Z_r	r	Z_r
0·01	0·0100	0·21	0·2132
0·02	0·0200	0·22	0·2237
0·03	0·0300	0·23	0·2342
0·04	0·0400	0·24	0·2448
0·05	0·0500	0·25	0·2554
0·06	0·0601	0·26	0·2661
0·07	0·0701	0·27	0·2769
0·08	0·0802	0·28	0·2877
0·09	0·0902	0·29	0·2986
0·10	0·1003	0·30	0·3095
0·11	0·1104	0·31	0·3205
0·12	0·1206	0·32	0·3316
0·13	0·1307	0·33	0·3428
0·14	0·1409	0·34	0·3541
0·15	0·1511	0·35	0·3654
0·16	0·1614	0·36	0·3769
0·17	0·1717	0·37	0·3884
0·18	0·1820	0·38	0·4001
0·19	0·1923	0·39	0·4118
0·20	0·2027	0·40	0·4236

Table III Fisher's transform of Pearson's correlation coefficient (continued)

r	Z_r	r	Z_r
0·41	0·4356	0·71	0·8872
0·42	0·4477	0·72	0·9076
0·43	0·4599	0·73	0·9287
0·44	0·4722	0·74	0·9505
0·45	0·4847	0·75	0·9730
0·46	0·4973	0·76	0·9962
0·47	0·5101	0·77	1·0203
0·48	0·5230	0·78	1·0454
0·49	0·5361	0·79	1·0714
0·50	0·5493	0·80	1·0986
0·51	0·5627	0·81	1·1270
0·52	0·5763	0·82	1·1568
0·53	0·5901	0·83	1·1881
0·54	0·6042	0·84	1·2212
0·55	0·6184	0·85	1·2562
0·56	0·6328	0·86	1·2933
0·57	0·6475	0·87	1·3331
0·58	0·6625	0·88	1·3758
0·59	0·6777	0·89	1·4219
0·60	0·6931	0·90	1·4722
0·61	0·7089	0·91	1·5275
0·62	0·7250	0·92	1·5890
0·63	0·7414	0·93	1·6584
0·64	0·7482	0·94	1·7380
0·65	0·7753	0·95	1·8318
0·66	0·7928	0·96	1·9459
0·67	0·8107	0·97	2·0923
0·68	0·8291	0·98	2·2976
0·69	0·8480	0·99	2·6467
0·70	0·8673		

Table IV The 5 (roman type) and 1 (boldface type) per cent points for the distribution of F

n_2	n_1 degrees of freedom (for greater mean square)																							
	1	2	3	4	5	6	7	8	9	10	11	12	14	16	20	24	30	40	50	75	100	200	500	∞
1	161	200	216	225	230	234	237	239	241	242	243	244	245	246	248	249	250	251	252	253	253	254	254	254
	4,052	**4,999**	**5,403**	**5,625**	**5,764**	**5,859**	**5,928**	**5,981**	**6,022**	**6,056**	**6,082**	**6,106**	**6,142**	**6,169**	**6,208**	**6,234**	**6,258**	**6,286**	**6,302**	**6,323**	**6,334**	**6,352**	**6,361**	**6,366**
2	18·51	19·00	19·16	19·25	19·30	19·33	19·36	19·37	19·38	19·39	19·40	19·41	19·42	19·43	19·44	19·45	19·46	19·47	19·47	19·48	19·49	19·49	19·50	19·50
	98·49	**99·00**	**99·17**	**99·25**	**99·30**	**99·33**	**99·34**	**99·36**	**99·38**	**99·40**	**99·41**	**99·42**	**99·43**	**99·44**	**99·45**	**99·46**	**99·47**	**99·48**	**99·48**	**99·49**	**99·49**	**99·49**	**99·50**	**99·50**
3	10·13	9·55	9·28	9·12	9·01	8·94	8·88	8·84	8·81	8·78	8·76	8·74	8·71	8·69	8·66	8·64	8·62	8·60	8·58	8·57	8·56	8·54	8·54	8·53
	34·12	**30·82**	**29·46**	**28·71**	**28·24**	**27·91**	**27·67**	**27·49**	**27·34**	**27·23**	**27·13**	**27·05**	**26·92**	**26·83**	**26·69**	**26·60**	**26·50**	**26·41**	**26·35**	**26·27**	**26·23**	**26·18**	**26·14**	**26·12**
4	7·71	6·94	6·59	6·39	6·26	6·16	6·09	6·04	6·00	5·96	5·93	5·91	5·87	5·84	5·80	5·77	5·74	5·71	5·70	5·68	5·66	5·65	5·64	5·63
	21·20	**18·00**	**16·69**	**15·98**	**15·52**	**15·21**	**14·98**	**14·80**	**14·66**	**14·54**	**14·45**	**14·37**	**14·24**	**14·15**	**14·02**	**13·93**	**13·83**	**13·74**	**13·69**	**13·61**	**13·57**	**13·52**	**13·48**	**13·46**
5	6·61	5·79	5·41	5·19	5·05	4·95	4·88	4·82	4·78	4·74	4·70	4·68	4·64	4·60	4·56	4·53	4·50	4·46	4·44	4·42	4·40	4·38	4·37	4·36
	16·26	**13·27**	**12·06**	**11·39**	**10·97**	**10·67**	**10·45**	**10·27**	**10·15**	**10·05**	**9·96**	**9·89**	**9·77**	**9·68**	**9·55**	**9·47**	**9·38**	**9·29**	**9·24**	**9·17**	**9·13**	**9·07**	**9·04**	**9·02**
6	5·99	5·14	4·76	4·53	4·39	4·28	4·21	4·15	4·10	4·06	4·03	4·00	3·96	3·92	3·87	3·84	3·81	3·77	3·75	3·72	3·71	3·69	3·68	3·67
	13·74	**10·92**	**9·78**	**9·15**	**8·75**	**8·47**	**8·26**	**8·10**	**7·98**	**7·87**	**7·79**	**7·72**	**7·60**	**7·52**	**7·39**	**7·31**	**7·23**	**7·14**	**7·09**	**7·02**	**6·99**	**6·94**	**6·90**	**6·88**
7	5·59	4·74	4·35	4·12	3·97	3·87	3·79	3·73	3·68	3·63	3·60	3·57	3·52	3·49	3·44	3·41	3·38	3·34	3·32	3·29	3·28	3·25	3·24	3·23
	12·25	**9·55**	**8·45**	**7·85**	**7·46**	**7·19**	**7·00**	**6·84**	**6·71**	**6·62**	**6·54**	**6·47**	**6·35**	**6·27**	**6·15**	**6·07**	**5·98**	**5·90**	**5·85**	**5·78**	**5·75**	**5·70**	**5·67**	**5·65**
8	5·32	4·46	4·07	3·84	3·69	3·58	3·50	3·44	3·39	3·34	3·31	3·28	3·23	3·20	3·15	3·12	3·08	3·05	3·03	3·00	2·98	2·96	2·94	2·93
	11·26	**8·65**	**7·59**	**7·01**	**6·63**	**6·37**	**6·19**	**6·03**	**5·91**	**5·82**	**5·74**	**5·67**	**5·56**	**5·48**	**5·36**	**5·28**	**5·20**	**5·11**	**5·06**	**5·00**	**4·96**	**4·91**	**4·88**	**4·86**
9	5·12	4·26	3·86	3·63	3·48	3·37	3·29	3·23	3·18	3·13	3·10	3·07	3·02	2·98	2·93	2·90	2·86	2·82	2·80	2·77	2·76	2·73	2·72	2·71
	10·56	**8·02**	**6·99**	**6·42**	**6·06**	**5·80**	**5·62**	**5·47**	**5·35**	**5·26**	**5·18**	**5·11**	**5·00**	**4·92**	**4·80**	**4·73**	**4·64**	**4·56**	**4·51**	**4·45**	**4·41**	**4·36**	**4·33**	**4·31**
10	4·96	4·10	3·71	3·48	3·33	3·22	3·14	3·07	3·02	2·97	2·94	2·91	2·86	2·82	2·77	2·74	2·70	2·67	2·64	2·61	2·59	2·56	2·55	2·54
	10·04	**7·56**	**6·55**	**5·99**	**5·64**	**5·39**	**5·21**	**5·06**	**4·95**	**4·85**	**4·78**	**4·71**	**4·60**	**4·52**	**4·41**	**4·33**	**4·25**	**4·17**	**4·12**	**4·05**	**4·01**	**3·96**	**3·93**	**3·91**
11	4·84	3·98	3·59	3·36	3·20	3·09	3·01	2·95	2·90	2·86	2·82	2·79	2·74	2·70	2·65	2·61	2·57	2·53	2·50	2·47	2·45	2·42	2·41	2·40
	9·65	**7·20**	**6·22**	**5·67**	**5·32**	**5·07**	**4·88**	**4·74**	**4·63**	**4·54**	**4·46**	**4·40**	**4·29**	**4·21**	**4·10**	**4·02**	**3·94**	**3·86**	**3·80**	**3·74**	**3·70**	**3·66**	**3·62**	**3·60**
12	4·75	3·88	3·49	3·26	3·11	3·00	2·92	2·85	2·80	2·76	2·72	2·69	2·64	2·60	2·54	2·50	2·46	2·42	2·40	2·36	2·35	2·32	2·31	2·30
	9·33	**6·93**	**5·95**	**5·41**	**5·06**	**4·82**	**4·65**	**4·50**	**4·39**	**4·30**	**4·22**	**4·16**	**4·05**	**3·98**	**3·86**	**3·78**	**3·70**	**3·61**	**3·56**	**3·49**	**3·46**	**3·41**	**3·38**	**3·36**
13	4·67	3·80	3·41	3·18	3·02	2·92	2·84	2·77	2·72	2·67	2·63	2·60	2·55	2·51	2·46	2·42	2·38	2·34	2·32	2·28	2·26	2·24	2·22	2·21
	9·07	**6·70**	**5·74**	**5·20**	**4·86**	**4·62**	**4·44**	**4·30**	**4·19**	**4·10**	**4·02**	**3·96**	**3·85**	**3·78**	**3·67**	**3·59**	**3·51**	**3·42**	**3·37**	**3·30**	**3·27**	**3·21**	**3·18**	**3·16**

Table IV The 5 (roman type) and 1 (boldface type) per cent points for the distribution of *F* (continued)

n_2	n_1 degrees of freedom (for greater mean square)																						
	1	2	3	4	5	6	7	8	9	10	11	12	14	16	20	24	30	40	50	75	100	200	500	∞
14	4·60	3·74	3·34	3·11	2·96	2·85	2·77	2·70	2·65	2·60	2·56	2·53	2·48	2·44	2·39	2·35	2·31	2·27	2·24	2·21	2·19	2·16	2·14	2·13
	8·86	**6·51**	**5·56**	**5·03**	**4·69**	**4·46**	**4·28**	**4·14**	**4·03**	**3·94**	**3·86**	**3·80**	**3·70**	**3·62**	**3·51**	**3·43**	**3·34**	**3·26**	**3·21**	**3·14**	**3·11**	**3·06**	**3·02**	**3·00**
15	4·54	3·68	3·29	3·06	2·90	2·79	2·70	2·64	2·59	2·55	2·51	2·48	2·43	2·39	2·33	2·29	2·25	2·21	2·18	2·15	2·12	2·10	2·08	2·07
	8·68	**6·36**	**5·42**	**4·89**	**4·56**	**4·32**	**4·14**	**4·00**	**3·89**	**3·80**	**3·73**	**3·67**	**3·56**	**3·48**	**3·36**	**3·29**	**3·20**	**3·12**	**3·07**	**3·00**	**2·97**	**2·92**	**2·89**	**2·87**
16	4·49	3·63	3·24	3·01	2·85	2·74	2·66	2·59	2·54	2·49	2·45	2·42	2·37	2·33	2·28	2·24	2·20	2·16	2·13	2·09	2·07	2·04	2·02	2·01
	8·53	**6·23**	**5·29**	**4·77**	**4·44**	**4·20**	**4·03**	**3·89**	**3·78**	**3·69**	**3·61**	**3·55**	**3·45**	**3·37**	**3·25**	**3·18**	**3·10**	**3·01**	**2·96**	**2·89**	**2·86**	**2·80**	**2·77**	**2·75**
17	4·45	3·59	3·20	2·96	2·81	2·70	2·62	2·55	2·50	2·45	2·41	2·38	2·33	2·29	2·23	2·19	2·15	2·11	2·08	2·04	2·02	1·99	1·97	1·96
	8·40	**6·11**	**5·18**	**4·67**	**4·34**	**4·10**	**3·93**	**3·79**	**3·68**	**3·59**	**3·52**	**3·45**	**3·35**	**3·27**	**3·16**	**3·08**	**3·00**	**2·92**	**2·86**	**2·79**	**2·76**	**2·70**	**2·67**	**2·65**
18	4·41	3·55	3·16	2·93	2·77	2·66	2·58	2·51	2·46	2·41	2·37	2·34	2·29	2·25	2·19	2·15	2·11	2·07	2·04	2·00	1·98	1·95	1·93	1·92
	8·28	**6·01**	**5·09**	**4·58**	**4·25**	**4·01**	**3·85**	**3·71**	**3·60**	**3·51**	**3·44**	**3·37**	**3·27**	**3·19**	**3·07**	**3·00**	**2·91**	**2·83**	**2·78**	**2·71**	**2·68**	**2·62**	**2·59**	**2·57**
19	4·38	3·52	3·13	2·90	2·74	2·63	2·55	2·48	2·43	2·38	2·34	2·31	2·26	2·21	2·15	2·11	2·07	2·02	2·00	1·96	1·94	1·91	1·90	1·88
	8·18	**5·93**	**5·01**	**4·50**	**4·17**	**3·94**	**3·77**	**3·63**	**3·52**	**3·43**	**3·36**	**3·30**	**3·19**	**3·12**	**3·00**	**2·92**	**2·84**	**2·76**	**2·70**	**2·63**	**2·60**	**2·54**	**2·51**	**2·49**
20	4·35	3·49	3·10	2·87	2·71	2·60	2·52	2·45	2·40	2·35	2·31	2·28	2·23	2·18	2·12	2·08	2·04	1·99	1·96	1·92	1·90	1·87	1·85	1·84
	8·10	**5·85**	**4·94**	**4·43**	**4·10**	**3·87**	**3·71**	**3·56**	**3·45**	**3·37**	**3·30**	**3·23**	**3·13**	**3·05**	**2·94**	**2·86**	**2·77**	**2·69**	**2·63**	**2·56**	**2·53**	**2·47**	**2·44**	**2·42**
21	4·32	3·47	3·07	2·84	2·68	2·57	2·49	2·42	2·37	2·32	2·28	2·25	2·20	2·15	2·09	2·05	2·00	1·96	1·93	1·89	1·87	1·84	1·82	1·81
	8·02	**5·78**	**4·87**	**4·37**	**4·04**	**3·81**	**3·65**	**3·51**	**3·40**	**3·31**	**3·24**	**3·17**	**3·07**	**2·99**	**2·88**	**2·80**	**2·72**	**2·63**	**2·58**	**2·51**	**2·47**	**2·42**	**2·38**	**2·36**
22	4·30	3·44	3·05	2·82	2·66	2·55	2·47	2·40	2·35	2·30	2·26	2·23	2·18	2·13	2·07	2·03	1·98	1·93	1·91	1·87	1·84	1·81	1·80	1·78
	7·94	**5·72**	**4·82**	**4·31**	**3·99**	**3·76**	**3·59**	**3·45**	**3·35**	**3·26**	**3·18**	**3·12**	**3·02**	**2·94**	**2·83**	**2·75**	**2·67**	**2·58**	**2·53**	**2·46**	**2·42**	**2·37**	**2·33**	**2·31**
23	4·28	3·42	3·03	2·80	2·64	2·53	2·45	2·38	2·32	2·28	2·24	2·20	2·14	2·10	2·04	2·00	1·96	1·91	1·88	1·84	1·82	1·79	1·77	1·76
	7·88	**5·66**	**4·76**	**4·26**	**3·94**	**3·71**	**3·54**	**3·41**	**3·30**	**3·21**	**3·14**	**3·07**	**2·97**	**2·89**	**2·78**	**2·70**	**2·62**	**2·53**	**2·48**	**2·41**	**2·37**	**2·32**	**2·28**	**2·26**
24	4·26	3·40	3·01	2·78	2·62	2·51	2·43	2·36	2·30	2·26	2·22	2·18	2·13	2·09	2·02	1·98	1·94	1·89	1·86	1·82	1·80	1·76	1·74	1·73
	7·82	**5·61**	**4·72**	**4·22**	**3·90**	**3·67**	**3·50**	**3·36**	**3·25**	**3·17**	**3·09**	**3·03**	**2·93**	**2·85**	**2·74**	**2·66**	**2·58**	**2·49**	**2·44**	**2·36**	**2·33**	**2·27**	**2·23**	**2·21**
25	4·24	3·38	2·99	2·76	2·60	2·49	2·41	2·34	2·28	2·24	2·20	2·16	2·11	2·06	2·00	1·96	1·92	1·87	1·84	1·80	1·77	1·74	1·72	1·71
	7·77	**5·57**	**4·68**	**4·18**	**3·86**	**3·63**	**3·46**	**3·32**	**3·21**	**3·13**	**3·05**	**2·99**	**2·89**	**2·81**	**2·70**	**2·62**	**2·54**	**2·45**	**2·40**	**2·32**	**2·29**	**2·23**	**2·19**	**2·17**
26	4·22	3·37	2·98	2·74	2·59	2·47	2·39	2·32	2·27	2·22	2·18	2·15	2·10	2·05	1·99	1·95	1·90	1·85	1·82	1·78	1·76	1·72	1·70	1·69
	7·72	**5·53**	**4·64**	**4·14**	**3·82**	**3·59**	**3·42**	**3·29**	**3·17**	**3·09**	**3·02**	**2·96**	**2·86**	**2·77**	**2·66**	**2·58**	**2·50**	**2·41**	**2·36**	**2·28**	**2·25**	**2·19**	**2·15**	**2·13**

Table IV The 5 (roman type) and 1 (boldface type) per cent points for the distribution of *F* (continued)

n_2	n_1 degrees of freedom (for greater mean square)																							
	1	2	3	4	5	6	7	8	9	10	11	12	14	16	20	24	30	40	50	75	100	200	500	∞
27	4·21	3·35	2·96	2·73	2·57	2·46	2·37	2·30	2·25	2·20	2·16	2·13	2·08	2·03	1·97	1·93	1·88	1·84	1·80	1·76	1·74	1·71	1·68	1·67
	7·68	**5·49**	**4·60**	**4·11**	**3·79**	**3·56**	**3·39**	**3·26**	**3·14**	**3·06**	**2·98**	**2·93**	**2·83**	**2·74**	**2·63**	**2·55**	**2·47**	**2·38**	**2·33**	**2·25**	**2·21**	**2·16**	**2·12**	**2·10**
28	4·20	3·34	2·95	2·71	2·56	2·44	2·36	2·29	2·24	2·19	2·15	2·12	2·06	2·02	1·96	1·91	1·87	1·81	1·78	1·75	1·72	1·69	1·67	1·65
	7·64	**5·45**	**4·57**	**4·07**	**3·76**	**3·53**	**3·36**	**3·23**	**3·11**	**3·03**	**2·95**	**2·90**	**2·80**	**2·71**	**2·60**	**2·52**	**2·44**	**2·35**	**2·30**	**2·22**	**2·18**	**2·13**	**2·09**	**2·06**
29	4·18	3·33	2·93	2·70	2·54	2·43	2·35	2·28	2·22	2·18	2·14	2·10	2·05	2·00	1·94	1·90	1·85	1·80	1·77	1·73	1·71	1·68	1·65	1·64
	7·60	**5·42**	**4·54**	**4·04**	**3·73**	**3·50**	**3·33**	**3·20**	**3·08**	**3·00**	**2·92**	**2·87**	**2·77**	**2·68**	**2·57**	**2·49**	**2·41**	**2·32**	**2·27**	**2·19**	**2·15**	**2·10**	**2·06**	**2·03**
30	4·17	3·32	2·92	2·69	2·53	2·42	2·34	2·27	2·21	2·16	2·12	2·09	2·04	1·99	1·93	1·89	1·84	1·79	1·76	1·72	1·69	1·66	1·64	1·62
	7·56	**5·39**	**4·51**	**4·02**	**3·70**	**3·47**	**3·30**	**3·17**	**3·06**	**2·98**	**2·90**	**2·84**	**2·74**	**2·66**	**2·55**	**2·47**	**2·38**	**2·29**	**2·24**	**2·16**	**2·13**	**2·07**	**2·03**	**2·01**
32	4·15	3·30	2·90	2·67	2·51	2·40	2·32	2·25	2·19	2·14	2·10	2·07	2·02	1·97	1·91	1·86	1·82	1·76	1·74	1·69	1·67	1·64	1·61	1·59
	7·50	**5·34**	**4·46**	**3·97**	**3·66**	**3·42**	**3·25**	**3·12**	**3·01**	**2·94**	**2·86**	**2·80**	**2·70**	**2·62**	**2·51**	**2·42**	**2·34**	**2·25**	**2·20**	**2·12**	**2·08**	**2·02**	**1·98**	**1·96**
34	4·13	3·28	2·88	2·65	2·49	2·38	2·30	2·23	2·17	2·12	2·08	2·05	2·00	1·95	1·89	1·84	1·80	1·74	1·71	1·67	1·64	1·61	1·59	1·57
	7·44	**5·29**	**4·42**	**3·93**	**3·61**	**3·38**	**3·21**	**3·08**	**2·97**	**2·89**	**2·82**	**2·76**	**2·66**	**2·58**	**2·47**	**2·38**	**2·30**	**2·21**	**2·15**	**2·08**	**2·04**	**1·98**	**1·94**	**1·91**
36	4·11	3·26	2·86	2·63	2·48	2·36	2·28	2·21	2·15	2·10	2·06	2·03	1·98	1·93	1·87	1·82	1·78	1·72	1·69	1·65	1·62	1·59	1·56	1·55
	7·39	**5·25**	**4·38**	**3·89**	**3·58**	**3·35**	**3·18**	**3·04**	**2·94**	**2·86**	**2·78**	**2·72**	**2·62**	**2·54**	**2·43**	**2·35**	**2·26**	**2·17**	**2·12**	**2·04**	**2·00**	**1·94**	**1·90**	**1·87**
38	4·10	3·25	2·85	2·62	2·46	2·35	2·26	2·19	2·14	2·09	2·05	2·02	1·96	1·92	1·85	1·80	1·76	1·71	1·67	1·63	1·60	1·57	1·54	1·53
	7·35	**5·21**	**4·34**	**3·86**	**3·54**	**3·32**	**3·15**	**3·02**	**2·91**	**2·82**	**2·75**	**2·69**	**2·59**	**2·51**	**2·40**	**2·32**	**2·22**	**2·14**	**2·08**	**2·00**	**1·97**	**1·90**	**1·86**	**1·84**
40	4·08	3·23	2·84	2·61	2·45	2·34	2·25	2·18	2·12	2·07	2·04	2·00	1·95	1·90	1·84	1·79	1·74	1·69	1·66	1·61	1·59	1·55	1·53	1·51
	7·31	**5·18**	**4·31**	**3·83**	**3·51**	**3·29**	**3·12**	**2·99**	**2·88**	**2·80**	**2·73**	**2·66**	**2·56**	**2·49**	**2·37**	**2·29**	**2·20**	**2·11**	**2·05**	**1·97**	**1·94**	**1·88**	**1·84**	**1·81**
42	4·07	3·22	2·83	2·59	2·44	2·32	2·24	2·17	2·11	2·06	2·02	1·99	1·94	1·89	1·82	1·78	1·73	1·68	1·64	1·60	1·57	1·54	1·51	1·49
	7·27	**5·15**	**4·29**	**3·80**	**3·49**	**3·26**	**3·10**	**2·96**	**2·86**	**2·77**	**2·70**	**2·64**	**2·54**	**2·46**	**2·35**	**2·26**	**2·17**	**2·08**	**2·02**	**1·94**	**1·91**	**1·85**	**1·80**	**1·78**
44	4·06	3·21	2·82	2·58	2·43	2·31	2·23	2·16	2·10	2·05	2·01	1·98	1·92	1·88	1·81	1·76	1·72	1·66	1·63	1·58	1·56	1·52	1·50	1·48
	7·24	**5·12**	**4·26**	**3·78**	**3·46**	**3·24**	**3·07**	**2·94**	**2·84**	**2·75**	**2·68**	**2·62**	**2·52**	**2·44**	**2·32**	**2·24**	**2·15**	**2·06**	**2·00**	**1·92**	**1·88**	**1·82**	**1·78**	**1·75**
46	4·05	3·20	2·81	2·57	2·42	2·30	2·22	2·14	2·09	2·04	2·00	1·97	1·91	1·87	1·80	1·75	1·71	1·65	1·62	1·57	1·54	1·51	1·48	1·46
	7·21	**5·10**	**4·24**	**3·76**	**3·44**	**3·22**	**3·05**	**2·92**	**2·82**	**2·73**	**2·66**	**2·60**	**2·50**	**2·42**	**2·30**	**2·22**	**2·13**	**2·04**	**1·98**	**1·90**	**1·86**	**1·80**	**1·76**	**1·72**
48	4·04	3·19	2·80	2·56	2·41	2·30	2·21	2·14	2·08	2·03	1·99	1·96	1·90	1·86	1·79	1·74	1·70	1·64	1·61	1·56	1·53	1·50	1·47	1·45
	7·19	**5·08**	**4·22**	**3·74**	**3·42**	**3·20**	**3·04**	**2·90**	**2·80**	**2·71**	**2·64**	**2·58**	**2·48**	**2·40**	**2·28**	**2·20**	**2·11**	**2·02**	**1·96**	**1·88**	**1·84**	**1·78**	**1·73**	**1·70**

Table IV The 5 (roman type) and 1 (boldface type) per cent points for the distribution of *F* (concluded)

n_2	n_1 degrees of freedom (for greater mean square)																							
	1	2	3	4	5	6	7	8	9	10	11	12	14	16	20	24	30	40	50	75	100	200	500	∞
50	4·03	3·18	2·79	2·56	2·40	2·29	2·20	2·13	2·07	2·02	1·98	1·95	1·90	1·85	1·78	1·74	1·69	1·63	1·60	1·55	1·52	1·48	1·46	1·44
	7·17	**5·06**	**4·20**	**3·72**	**3·41**	**3·18**	**3·02**	**2·88**	**2·78**	**2·70**	**2·62**	**2·56**	**2·46**	**2·39**	**2·26**	**2·18**	**2·10**	**2·00**	**1·94**	**1·86**	**1·82**	**1·76**	**1·71**	**1·68**
55	4·02	3·17	2·78	2·54	2·38	2·27	2·18	2·11	2·05	2·00	1·97	1·93	1·88	1·83	1·76	1·72	1·67	1·61	1·58	1·52	1·50	1·46	1·43	1·41
	7·12	**5·01**	**4·16**	**3·68**	**3·37**	**3·15**	**2·98**	**2·85**	**2·75**	**2·66**	**2·59**	**2·53**	**2·43**	**2·35**	**2·23**	**2·15**	**2·06**	**1·96**	**1·90**	**1·82**	**1·78**	**1·71**	**1·66**	**1·64**
60	4·00	3·15	2·76	2·52	2·37	2·25	2·17	2·10	2·04	1·99	1·95	1·92	1·86	1·81	1·75	1·70	1·65	1·59	1·56	1·50	1·48	1·44	1·41	1·39
	7·08	**4·98**	**4·13**	**3·65**	**3·34**	**3·12**	**2·95**	**2·82**	**2·72**	**2·63**	**2·56**	**2·50**	**2·40**	**2·32**	**2·20**	**2·12**	**2·03**	**1·93**	**1·87**	**1·79**	**1·74**	**1·68**	**1·63**	**1·60**
65	3·99	3·14	2·75	2·51	2·36	2·24	2·15	2·08	2·02	1·98	1·94	1·90	1·85	1·80	1·73	1·68	1·63	1·57	1·54	1·49	1·46	1·42	1·39	1·37
	7·04	**4·95**	**4·10**	**3·62**	**3·31**	**3·09**	**2·93**	**2·79**	**2·70**	**2·61**	**2·54**	**2·47**	**2·37**	**2·30**	**2·18**	**2·09**	**2·00**	**1·90**	**1·84**	**1·76**	**1·71**	**1·64**	**1·60**	**1·56**
70	3·98	3·13	2·74	2·50	2·35	2·23	2·14	2·07	2·01	1·97	1·93	1·89	1·84	1·79	1·72	1·67	1·62	1·56	1·53	1·47	1·45	1·40	1·37	1·35
	7·01	**4·92**	**4·08**	**3·60**	**3·29**	**3·07**	**2·91**	**2·77**	**2·67**	**2·59**	**2·51**	**2·45**	**2·35**	**2·28**	**2·15**	**2·07**	**1·98**	**1·88**	**1·82**	**1·74**	**1·69**	**1·62**	**1·56**	**1·53**
80	3·96	3·11	2·72	2·48	2·33	2·21	2·12	2·05	1·99	1·95	1·91	1·88	1·82	1·77	1·70	1·65	1·60	1·54	1·51	1·45	1·42	1·38	1·35	1·32
	6·96	**4·88**	**4·04**	**3·56**	**3·25**	**3·04**	**2·87**	**2·74**	**2·64**	**2·55**	**2·48**	**2·41**	**2·32**	**2·24**	**2·11**	**2·03**	**1·94**	**1·84**	**1·78**	**1·70**	**1·65**	**1·57**	**1·52**	**1·49**
100	3·94	3·09	2·70	2·46	2·30	2·19	2·10	2·03	1·97	1·92	1·88	1·85	1·79	1·75	1·68	1·63	1·57	1·51	1·48	1·42	1·39	1·34	1·30	1·28
	6·90	**4·82**	**3·98**	**3·51**	**3·20**	**2·99**	**2·82**	**2·69**	**2·59**	**2·51**	**2·43**	**2·36**	**2·26**	**2·19**	**2·06**	**1·98**	**1·89**	**1·79**	**1·73**	**1·64**	**1·59**	**1·51**	**1·46**	**1·43**
125	3·92	3·07	2·68	2·44	2·29	2·17	2·08	2·01	1·95	1·90	1·86	1·83	1·77	1·72	1·65	1·60	1·55	1·49	1·45	1·39	1·36	1·31	1·27	1·25
	6·84	**4·78**	**3·94**	**3·47**	**3·17**	**2·95**	**2·79**	**2·65**	**2·56**	**2·47**	**2·40**	**2·33**	**2·23**	**2·15**	**2·03**	**1·94**	**1·85**	**1·75**	**1·68**	**1·59**	**1·54**	**1·46**	**1·40**	**1·37**
150	3·91	3·06	2·67	2·43	2·27	2·16	2·07	2·00	1·94	1·89	1·85	1·82	1·76	1·71	1·64	1·59	1·54	1·47	1·44	1·37	1·34	1·29	1·25	1·22
	6·81	**4·75**	**3·91**	**3·44**	**3·14**	**2·92**	**2·76**	**2·62**	**2·53**	**2·44**	**2·37**	**2·30**	**2·20**	**2·12**	**2·00**	**1·91**	**1·83**	**1·72**	**1·66**	**1·56**	**1·51**	**1·43**	**1·37**	**1·33**
200	3·89	3·04	2·65	2·41	2·26	2·14	2·05	1·98	1·92	1·87	1·83	1·80	1·74	1·69	1·62	1·57	1·52	1·45	1·42	1·35	1·32	1·26	1·22	1·19
	6·76	**4·71**	**3·88**	**3·41**	**3·11**	**2·90**	**2·73**	**2·60**	**2·50**	**2·41**	**2·34**	**2·28**	**2·17**	**2·09**	**1·97**	**1·88**	**1·79**	**1·69**	**1·62**	**1·53**	**1·48**	**1·39**	**1·33**	**1·28**
400	3·86	3·02	2·62	2·39	2·23	2·12	2·03	1·96	1·90	1·85	1·81	1·78	1·72	1·67	1·60	1·54	1·49	1·42	1·38	1·32	1·28	1·22	1·16	1·13
	6·70	**4·66**	**3·83**	**3·36**	**3·06**	**2·85**	**2·69**	**2·55**	**2·46**	**2·37**	**2·29**	**2·23**	**2·12**	**2·04**	**1·92**	**1·84**	**1·74**	**1·64**	**1·57**	**1·47**	**1·42**	**1·32**	**1·24**	**1·19**
1000	3·85	3·00	2·61	2·38	2·22	2·10	2·02	1·95	1·89	1·84	1·80	1·76	1·70	1·65	1·58	1·53	1·47	1·41	1·36	1·30	1·26	1·19	1·13	1·08
	6·66	**4·62**	**3·80**	**3·34**	**3·04**	**2·82**	**2·66**	**2·53**	**2·43**	**2·34**	**2·26**	**2·20**	**2·09**	**2·01**	**1·89**	**1·81**	**1·71**	**1·61**	**1·54**	**1·44**	**1·38**	**1·28**	**1·19**	**1·11**
∞	3·84	2·99	2·60	2·37	2·21	2·09	2·01	1·94	1·88	1·83	1·79	1·75	1·69	1·64	1·57	1·52	1·46	1·40	1·35	1·28	1·24	1·17	1·11	1·00
	6·64	**4·60**	**3·78**	**3·32**	**3·02**	**2·80**	**2·64**	**2·51**	**2·41**	**2·32**	**2·24**	**2·18**	**2·07**	**1·99**	**1·87**	**1·79**	**1·69**	**1·59**	**1·52**	**1·41**	**1·36**	**1·25**	**1·15**	**1·00**

Table V Probabilities associated with observed values of *H* in the Kruskal–Wallis test

Sample sizes				
n_1	n_2	n_3	H	P
2	1	1	2·7000	0·500
2	2	1	3·6000	0·200
2	2	2	4·5714	0·067
			3·7143	0·200
3	1	1	3·2000	0·300
3	2	1	4·2857	0·100
			3·8571	0·133
3	2	2	5·3572	0·029
			4·7143	0·048
			4·5000	0·067
			4·4643	0·105
3	3	1	5·1429	0·043
			4·5714	0·100
			4·0000	0·129
3	3	2	6·2500	0·011
			5·3611	0·032
			5·1389	0·061
			4·5556	0·100
			4·2500	0·121
3	3	3	7·2000	0·004
			6·4889	0·011
			5·6889	0·029
			5·6000	0·050
			5·0667	0·086
			4·6222	0·100
4	1	1	3·5714	0·200
4	2	1	4·8214	0·057
			4·5000	0·076
			4·0179	0·114
4	2	2	6·0000	0·014
			5·3333	0·033
			5·1250	0·052
			4·4583	0·100
			4·1667	0·105
4	3	1	5·8333	0·021
			5·2083	0·050
			5·0000	0·057
			4·0556	0·093
			3·8889	0·129
4	3	2	6·4444	0·008
			6·3000	0·011
			5·4444	0·046
			5·4000	0·051
			4·5111	0·098
			4·4444	0·102
4	3	3	6·7455	0·010
			6·7091	0·013
			5·7909	0·046
			5·7273	0·050
			4·7091	0·092
			4·7000	0·101
4	4	1	6·6667	0·010
			6·1667	0·022
			4·9667	0·048
			4·8667	0·054
			4·1667	0·082
			4·0667	0·102
4	4	2	7·0364	0·006
			6·8727	0·011
			5·4545	0·046
			5·2364	0·052
			4·5545	0·098
			4·4455	0·103

Table V Probabilities associated with observed values of *H* in the Kruskal–Wallis test (continued)

Sample sizes n_1	n_2	n_3	H	P
4	4	3	7·1439	0·010
			7·1364	0·011
			5·5985	0·049
			5·5758	0·051
			4·5455	0·099
			4·4773	0·102
4	4	4	7·6538	0·008
			7·5385	0·011
			5·6923	0·049
			5·6538	0·054
			4·6539	0·097
			4·5001	0·104
5	1	1	3·8571	0·143
5	2	1	5·2500	0·036
			5·0000	0·048
			4·4500	0·071
			4·2000	0·095
			4·0500	0·119
			5·6485	0·049
5	2	2	6·5333	0·008
			6·1333	0·013
			5·1600	0·034
			5·0400	0·056
			4·3733	0·090
			4·2933	0·122
5	3	1	6·4000	0·012
			4·9600	0·048
			4·8711	0·052
			4·0178	0·095
			3·8400	0·123
5	3	2	6·9091	0·009
			6·8218	0·010
			5·2509	0·049
			5·1055	0·052
			4·6509	0·091
			4·4945	0·101
5	3	3	7·0788	0·009
			6·9818	0·011
			5·6485	0·049
			5·5152	0·051
			4·5333	0·097
			4·4121	0·109
5	4	1	6·9545	0·008
			6·8400	0·011
			4·9855	0·044
			4·8600	0·056
			3·9873	0·098
			3·9600	0·102
5	4	2	7·2045	0·009
			7·1182	0·010
			5·2727	0·049
			5·2682	0·050
			4·5409	0·098
			4·5182	0·101
5	4	3	7·4449	0·010
			7·3949	0·011
			5·6564	0·049
			5·6308	0·050
			4·5487	0·099
			4·5231	0·103
5	4	4	7·7604	0·009
			7·7440	0·011
			5·6571	0·049
			5·6176	0·050
			4·6187	0·100
			4·5527	0·102

Table V Probabilities associated with observed values of *H* in the Kruskal–Wallis test (concluded)

Sample sizes			*H*	*P*
n_1	n_2	n_3		
5	5	1	7·3091	0·009
			6·8364	0·011
			5·1273	0·046
			4·9091	0·053
			4·1091	0·086
			4·0364	0·105
5	5	2	7·3385	0·010
			7·2692	0·010
			5·3385	0·047
			5·2462	0·051
			4·6231	0·097
			4·5077	0·100
5	5	3	7·5780	0·010
			7·5429	0·010
			5·7055	0·046
			5·6264	0·051
			4·5451	0·100
			4·5363	0·102
5	5	4	7·8229	0·010
			7·7914	0·010
			5·6657	0·049
			5·6429	0·050
			4·5229	0·099
			4·5200	0·101
5	5	5	8·0000	0·009
			7·9800	0·010
			5·7800	0·049
			5·6600	0·051
			4·5600	0·100
			4·5000	0·102

From Kruskal, W. H., and Wallis, W. A. Use of ranks in one-criterion variance analysis. *Journal of the American Statistical Association*, 1952, *48*, 910. Copyright 1952 by the American Statistical Association. Reprinted by permission.

Table VI Critical values of χ^2

Degrees of freedom DF	0·05	0·02	0·01
1	3·841	5·412	6·635
2	5·991	7·824	9·210
3	7·815	9·837	11·341
4	9·488	11·668	13·277
5	11·070	13·388	15·086
6	12·592	15·033	16·812
7	14·067	16·622	18·475
8	15·507	18·168	20·090
9	16·919	19·679	21·666
10	18·307	21·161	23·209
11	19·675	22·618	24·725
12	21·026	24·054	26·217
13	22·362	25·472	27·688
14	23·685	26·873	29·141
15	24·996	28·259	30·578
16	26·296	29·633	32·000
17	27·587	30·995	33·409
18	28·869	32·346	34·805
19	30·144	33·687	36·191
20	31·410	35·020	37·566
21	32·671	36·343	38·932
22	33·924	37·659	40·289
23	35·172	38·968	41·638
24	36·415	40·270	42·980
25	37·652	41·566	44·314
26	38·885	42·856	45·642
27	40·113	44·140	46·963
28	41·337	45·419	48·278
29	42·557	46·693	49·588
30	43·773	47·962	50·892

For larger values of DF, the expression $\sqrt{2\chi^2} - \sqrt{2(DF) - 1}$ may be used as a normal deviate with unit standard error.

Table VII Probabilities associated with observed values of χ_r^2 in the Friedman test

Table N_I. $k = 3$

$N = 2$		$N = 3$		$N = 4$		$N = 5$	
χ_r^2	P	χ_r^2	P	χ_r^2	P	χ_r^2	P
0	1·000	0·000	1·000	0·0	1·000	0·0	1·000
1	0·833	0·667	0·944	0·5	0·931	0·4	0·954
3	0·500	2·000	0·528	1·5	0·653	1·2	0·691
4	0·167	2·667	0·361	2·0	0·431	1·6	0·522
		4·667	0·194	3·5	0·273	2·8	0·367
		6·000	0·028	4·5	0·125	3·6	0·182
				6·0	0·069	4·8	0·124
				6·5	0·042	5·2	0·093
				8·0	0·0046	6·4	0·039
						7·6	0·024
						8·4	0·0085
						10·0	0·00077

$N = 6$		$N = 7$		$N = 8$		$N = 9$	
χ_r^2	P	χ_r^2	P	χ_r^2	P	χ_r^2	P
0·00	1·000	0·000	1·000	0·00	1·000	0·000	1·000
0·33	0·956	0·286	0·964	0·25	0·967	0·222	0·971
1·00	0·740	0·857	0·768	0·75	0·794	0·667	0·814
1·33	0·570	1·143	0·620	1·00	0·654	0·889	0·865
2·33	0·430	2·000	0·486	1·75	0·531	1·556	0·569
3·00	0·252	2·571	0·305	2·25	0·355	2·000	0·398
4·00	0·184	3·429	0·237	3·00	0·285	2·667	0·328
4·33	0·142	3·714	0·192	3·25	0·236	2·889	0·278
5·33	0·072	4·571	0·112	4·00	0·149	3·556	0·187
6·33	0·052	5·429	0·085	4·75	0·120	4·222	0·154
7·00	0·029	6·000	0·052	5·25	0·079	4·667	0·107
8·33	0·012	7·143	0·027	6·25	0·047	5·556	0·069
9·00	0·0081	7·714	0·021	6·75	0·038	6·000	0·057
9·33	0·0055	8·000	0·016	7·00	0·030	6·222	0·048
10·33	0·0017	8·857	0·0084	7·75	0·018	6·889	0·031
12·00	0·00013	10·286	0·0036	9·00	0·0099	8·000	0·019
		10·571	0·0027	9·25	0·0080	8·222	0·016
		11·143	0·0012	9·75	0·0048	8·667	0·010
		12·286	0·00032	10·75	0·0024	9·556	0·0060
		14·000	0·000021	12·00	0·0011	10·667	0·0035
				12·25	0·00086	10·889	0·0029
				13·00	0·00026	11·556	0·0013
				14·25	0·000061	12·667	0·00066
				16·00	0·0000036	13·556	0·00035
						14·000	0·00020
						14·222	0·000097
						14·889	0·000054
						16·222	0·000011
						18·000	0·0000006

Table VII Probabilities associated with observed values of χ_r^2 in the Friedman test (continued)

Table N_{II}. $k = 4$

$N = 2$		$N = 3$		$N = 4$			
χ_r^2	P	χ_r^2	P	χ_r^2	P	χ_r^2	P
0·0	1·000	0·2	1·000	0·0	1·000	5·7	0·141
0·6	0·958	0·6	0·958	0·3	0·992	6·0	0·105
1·2	0·834	1·0	0·910	0·6	0·928	6·3	0·094
1·8	0·792	1·8	0·727	0·9	0·900	6·6	0·077
2·4	0·625	2·2	0·608	1·2	0·800	6·9	0·068
3·0	0·542	2·6	0·524	1·5	0·754	7·2	0·054
3·6	0·458	3·4	0·446	1·8	0·677	7·5	0·052
4·2	0·375	3·8	0·342	2·1	0·649	7·8	0·036
4·8	0·208	4·2	0·300	2·4	0·524	8·1	0·033
5·4	0·167	5·0	0·207	2·7	0·508	8·4	0·019
6·0	0·042	5·4	0·175	3·0	0·432	8·7	0·014
		5·8	0·148	3·3	0·389	9·3	0·012
		6·6	0·075	3·6	0·355	9·6	0·0069
		7·0	0·054	3·9	0·324	9·9	0·0062
		7·4	0·033	4·5	0·242	10·2	0·0027
		8·2	0·017	4·8	0·200	10·8	0·0016
		9·0	0·0017	5·1	0·190	11·1	0·00094
				5·4	0·158	12·0	0·000072

From Friedman, M. The use of ranks to avoid assumptions of normality implicit in analysis of variance. *Journal of the American Statistical Association*, 1937, 32, 688–689. Copyright 1937 by the American Statistical Association. Reprinted by permission.

Table VIII Critical values of *L* in Page's trend test

2-tail 1-tail	10% 5%	5% 2·5%	2% 1%	0·2% 0·1%
$k = 3$ $n = 2$	$L \geqslant 28$			
3	41	42	42	
4	54	55	55	56
5	67	68	68	70
6	79	80	81	83
$k = 4$ $n = 2$	58	59	60	
3	84	86	87	89
4	111	112	114	117
5	137	138	140	145

Reprinted by permission from R. Meddis, *Statistical Handbook for Non-Statisticians*, copyright © 1975, McGraw-Hill, Maidenhead, England.

References

Adinolfi, A. A. Relevance of person perception to clinical psychology. *Journal of Consulting and Clinical Psychology*, 1971, *37*, 167–176.

Ayton, P., and Wright, G. Persons, situations, interactions and error: consistency, variability and confusion. *Personality and individual differences*, in press.

Barber, T. X. *Pitfalls in human research: ten pivotal points*. New York: Pergamon, 1976.

Barnes, J. A. *Who should know what? Social science, privacy and ethics*. London: Penguin, 1979.

Binder, A., McConnell, E., and Sjoholm, N. A. Verbal conditioning as a function of experimenter characteristics. *Journal of Abnormal and Social Psychology*, 1957, *55*, 309–14.

Bindra, D., and Schier, I. The relation between psychometric and experimental research in psychology. *American Psychologist*, 1954, *9*, 69–71.

Blakemore, C. Environmental constraints on development in the visual system. In R. A. Hinde and J. Stevenson-Hinde (eds.), *Constraints on learning*. London, Academic Press, 1973.

Bowers, K. S. Situationism in psychology. An analysis and a critique. *Psychological Review*, 1973, *80*, 309–336.

Carmines, E. G., and Zeller, R. A. *Reliability and validity assessment*. Beverly Hills: Sage Publications, 1979.

Cattell, R. B. *The description and measurement of personality*. New York: World Books, 1946.

Child, D. *The essentials of factor analysis*. London: Holt, Rinehart & Winston, 1977.

Claxton, G. Cognitive psychology: a suitable case for what sort of treatment. In G. Claxton (ed.), *Cognitive psychology: new directions*. London: Routledge & Kegan Paul, 1980.

Comrey, A. L. *A first course in factor analysis*. New York: Academic Press, 1973.

Cordaro, L., and Ison, J. R. Observer bias in classical conditioning of the planarian. *Psychology Reports*, 1963, *13*, 787–789.

Cronbach, L. J. The two disciplines of scientific psychology. *American Psychologist*, 1957, *12*, 671–684.

Cronbach, L. J. Beyond the two disciplines of scientific psychology. *American Psychologist*, 1975, *30*, 116–127.

Cronbach, L. J., Glaser, G. C., Nanda, H., and Rajaratnam, N. *The dependability of behavioural measurement: Theory of generalisability for scores and profiles*. New York: Wiley, 1972.

Dawes, R. M. Graduate admission variables and future success. *Science*, 1975, *187*, 721–743.

Dawes, R. M., and Corrigan, B. Linear models in decision-making. *Psychological Bulletin*, 1974, *81*, 95–106.

Denzin, N. K. (ed.). *Sociological methods: A sourcebook*. Chicago: Aldine, 1970.

Einhorn, H. J. Expert measurement and mechanical combination. *Organizational Behavior and Human Performance*, 1972, 7, 86–106.

Endler, N. S. Estimating variance components from mean squares for random and mixed effects analysis of variance models. *Perceptual and Motor Skills*, 1966, 22, 559–570.

Endler, N. S. The case for person-situation interactions. *Canadian Psychological Review*, 1975, *16*, 319–329.

Endler, N. S., and Hunt, J. McV. S–R inventories of hostility and comparisons of the proportions of variance from persons, responses and situations for hostility and anxiousness. *Journal of Personality and Social Psychology*, 1968, *9*, 309–315.

Folkard, S. Circadian rhythms and human memory. In F. M. Brown and R. Curtis Graeber (eds.), *Rhythmic aspects of behavior*, New Jersey: Erlbaum, 1982.

Forcese, D. P., and Richer, S. *Social research methods*. Prentice-Hall: New Jersey, 1973.

Friedman, N. *The social nature of psychological research*. New York: Basic Books, 1967.

Gaito, J. Measurement scales and statistics: Resurgence of an old misconception. *Psychological Bulletin*, 1980, *87*, 564–567.

Goldberg, J. R. Diagnosticians versus diagnostic signs of the disgnosis of psychosis versus neurosis from the MMPI. *Psychological Monographs*, 1965, *79*, 602–643.

Golding, S. L. Flies in the ointment: Methodological problems in the analysis of percentage of variance due to persons and situations. *Psychological Bulletin*, 1975, *82*, 278–288.

Goodman, L. D., and Kruskal, W. H. Measures of association for cross-classifications. *Journal of the American Statistical Association*, 1954, *49*, 732–764.

Guildford, J. P. *Personality*. New York: McGraw-Hill, 1959.

Hartley, H. O. The maximum *F*-ratio as a short-cut test for heterogeneity of variance. *Biometrika*, 1950, *37*, 308–312.

Hartwig, F., and Dearing, A. *Exploratory Data Analysis*. Beverly Hills: Sage Publications, 1979.

Hays, W. L. *Statistics*. New York: Holt, Rinehart & Winston, 1963.

Hays, W. L. *Statistics for the social sciences*. New York: Holt, Rinehart & Winston, 1974.

Heerman, E. F., and Braskamp, L. A. (eds.). Readings in statistics for the behavioural sciences. Englewood Cliffs, N.J.: Prentice-Hall, 1970.

Howell, D. *Statistical methods for psychology*. Boston: PWS, 1982.

Isen, A. M. Success, failure, attention and reaction to others. The warm glow of success. *Journal of Personality and Social Psychology*, 1970, *15*, 294–301.

Johnston, J. M., and Pennypacker, H. S. *Strategies and tactics of human behavioural research*. New Jersey: Erlbaum, 1980.

Kerlinger, F. W. *Foundations of behavioral research*. New York: Holt, Rinehart & Winston, 1973.

Kuhn, T. S. The structure of scientific revolutions. Chicago: University of Chicago Press, 1970.

Labov, W. The logical of non-standard English. In F. Williams, *Language and poverty*, Chicago: Markham, 1970.

Latané, B., and Darley, J. M. *The unresponsive bystander: why doesn't he help?* New York: Appleton-Century-Crofts, 1970.

Lindquist, E. F. *Design and analysis of experiments in psychology and education.* Boston: Houghton Mifflin, 1953.

McClelland, D. C. *Personality.* New York: Dryden, 1951.

Mason, E. J., et al. Three approaches to teaching and learning in education: Behavioural, Piagetian and information-processing. *International Science*, 1983, 12, 219–241.

Meehl, P. E. A comparison of clinicians with five statistical methods of identifying psychotic MMPI profiles. *Journal of Counselling Psychology*, 1969, *6*, 102–122.

Milgram, S. Behavioural studies of obedience. *Journal of Abnormal and Social Psychology*, 1963, *67*, 371–378.

Miller, A. G. Role playing: an alternative to deception? *American Psychology*, 1972, *27*, 623–636.

Mischel, W. *Personality and assessment.* New York: Wiley, 1968.

Nisbett, R. E. Interaction versus main effect as goal of personality research. In D. Magnusson and N. S. Endler (eds.), *Personality at the crossroads.* New York: Lawrence Erlbaum Associates, 1977.

Olweus, D. A critical analysis of the 'modern' interactionist position. In D. Magnusson and N. S. Endler (eds.), *Personality at the crossroads.* New York: Lawrence Erlbaum Associates, 1977.

Orne, M. T. On the social psychology of the psychology experiment with particular reference to demand characteristics and their implications. *American Psychologist*, 1962, *17*, 776–783.

Payne, J. W. Contingent decision behavior. *Psychological Bulletin*, 1982, *92*, 382–402.

Plutchik, R. *Foundations of experimental research.* New York: Harper Row, 1983.

Resnick, J. H., and Schwartz, T. Ethical standards as an independent variable in psychological research. *American Psychologist*, 1973, *28*, 134–139.

Robson, C. *Experiment, design and statistics in psychology.* Harmondsworth: Penguin, 1983.

Rosenberg, M. J. The conditions and consequences of evaluation apprehension. In R. Rosenthal and R. L. Rosnow (eds.), *Artifacts in behavioral research.* New York: Academic Press, 1969.

Rosenthal, R. *Experimental effects in behavioral research.* New York: Irvington, 1976.

Rosenthal, R., and Rosnow, R. L. *The volunteer subject.* New York: Wiley, 1975.

Shavelson, R. J. *Statistical reasons for the behavioral sciences.* Boston: Allyn & Bacon, 1981.

Siegel, S. *Nonparametric statistics for behavioral sciences.* New York: McGraw-Hill, 1956.

Spector, P. E. *Research designs.* Beverly Hills: Sage Publications, 1981.

Tukey, J. W. *Exploratory data analysis.* Reading, Mass. Addison-Wesley, 1977.

Whalley, P. C. *'The psychology of similarity'.* Unpublished PhD thesis, Open University, 1984.

Winer, B. J. *Statistical principles in experimental design.* New York: McGraw-Hill, 1971.

Index

MORE ABOUT PENGUINS, PELICANS AND PUFFINS